AN INTRODUCTION TO THE KINETIC THEORY OF GASES

AN INTRODUCTION TO
THE KINETIC THEORY
OF GASES

BY

SIR JAMES JEANS

CAMBRIDGE
AT THE UNIVERSITY PRESS
1952

PUBLISHED BY
THE SYNDICS OF THE CAMBRIDGE UNIVERSITY PRESS

London Office: Bentley House, N.W. I
American Branch: New York

Agents for Canada, India, and Pakistan: Macmillan

First Edition 1940
Reprinted 1946
1948
1952

First printed in Great Britain at The University Press, Cambridge
Reprinted by Spottiswoode, Ballantyne & Co., Ltd., Colchester

PREFACE

I have intended that the present book shall provide such knowledge of the Kinetic Theory as is required by the average serious student of physics and physical chemistry. I hope it will also give the mathematical student the equipment he should have before undertaking the study of specialist monographs, such, for instance, as the recent books of Chapman and Cowling (*The Mathematical Theory of Non-uniform Gases*) and R. H. Fowler (*Statistical Thermodynamics*).

Inevitably the book covers a good deal of the same ground as my earlier book, *The Dynamical Theory of Gases*, but it is covered in a simpler and more physical manner. Primarily I have kept before me the physicist's need for clearness and directness of treatment rather than the mathematician's need for rigorous general proofs. This does not mean that many subjects will not be found treated in the same way—and often in the same words—in the two books; I have tried to retain all that was of physical interest in the old book, while discarding much of which the interest was mainly mathematical.

It is a pleasure to thank Professor E. N. da C. Andrade for reading my proofs, and suggesting many improvements which have greatly enhanced the value of the book. I am also greatly indebted to W. F. Sedgwick,* sometime of Trinity College, Cambridge, for checking all the numerical calculations in the latest edition of my old book, and suggesting many improvements.

J. H. JEANS

DORKING
June 1940

[* W. F. Sedgwick writes (1946): "As a rule I only checked one or two of the items in the tables. As regards these and the numerical results given in the text, I did indeed as a rule agree, at least approximately, with Jeans' figures, but in a few cases (see *Philosophical Magazine*, Sept. 1946, p. 651) my results differed substantially."

PUBLISHERS' NOTE.]

CONTENTS

Chapter I

INTRODUCTION

The Origins of the Theory

1. As soon as man began to think of abstract problems at all, it was only natural that speculations as to the nature and ultimate structure of the material world should figure largely in his writings and philosophies.

Among the earliest speculations which have survived are those of Thales of Miletus (about 640–547 B.C.), many of whose ideas may well have been derived from still earlier legends of Egyptian origin. He conjectured that the whole material universe consisted only of water and of substances derived from water by physical transformation. Earth was produced by the condensation of water, and air by its rarefaction, while air when heated became fire. About 500 B.C. Heraclitus advanced the alternative view that earth, air, fire and water were not transformable one into the other, but constituted four distinct unalterable "elements", and that all material substances were composed of these four elements mixed in varying proportions—a sort of dim anticipation of modern chemical theory. At a somewhat later date, Leucippus and Democritus maintained that matter consisted of minute hard particles moving as separate units in empty space, and that there were as many kinds of particles as there are different substances.

Unhappily nothing now remains of the writings of either Democritus or Leucippus; their opinions are known to us only through second-hand accounts. From these we learn that they imagined their particles to be eternal and invisible, and so small that their size could not be diminished; hence the name ἄτομος—indivisible. The particles of any particular substance, such as iron or water, were supposed to be all similar to one another, and every one of them carried in itself all the attributes of the substance. For instance, Democritus taught that the atoms of water, being smooth and round, are unable to hook on to each other, so that they roll over and over like small globes; on the other hand, the atoms of iron, being rough, jagged and uneven, cling together to form a solid body. The "atoms" of the Greeks corresponded of course to the molecules of modern chemistry.

Similar views were advocated by Epicurus (341–276 B.C.), but rejected by Aristotle. In a later age (A.D. 55) Lucretius advanced substantially identical ideas in his great poem *De rerum natura*. In this he claimed only to expound the views of Epicurus, most of whose writings are now lost.

Lucretius explains very clearly that the atoms of all bodies are in ceaseless motion, colliding and rebounding from one another. When the distances which the atoms cover between successive rebounds are small, the substance is in the solid state; when large, we have "thin air and bright sunshine". He further explains that atoms must be very small, as can be seen either from the imperceptible wearing away of objects, or from the way in which our clothes can become damp without exhibiting visible drops of moisture.*

There was little further discussion of the problem until the middle of the seventeenth century, when Gassendi† examined some of the physical consequences of the atomic view. He assumed his atoms to be similar in substance, although different in size and form, to move in all directions through empty space, and to be devoid of all qualities except absolute rigidity. With these simple assumptions, Gassendi was able to explain a number of physical phenomena, including the three states of matter and the transitions from one to another, in a way which differed but little from that of the modern kinetic theory. He further saw that in an ordinary gas, such as atmospheric air, the particles must be very widely spaced, and he was, so far as we know, the first to conjecture that the motion of these particles could account by itself for a number of well-known physical phenomena, without the addition of separate *ad hoc* hypotheses. All this gives him a very special claim to be regarded as the father of the kinetic theory.

* See an essay by E. N. da C. Andrade, "The Scientific Significance of Lucretius", Munro's *Lucretius*, 4th edition (Bell, 1928).

† *Syntagma Philosophicum*, 1658, Lugduni.

Twenty years after Gassendi, Hooke advanced somewhat similar ideas. He suggested that the elasticity of a gas resulted from the impact of hard independent particles on the substance which enclosed it, and even tried to explain Boyle's law on this basis.

Newton accepted these views as to the atomic structure of matter, although he suggested a different explanation of Boyle's law*. He wrote†

"It seems probable to me that God in the beginning formed matter in solid, massy, hard, impenetrable, moveable particles ..., and that these primary particles, being solids, are incomparably harder than any porous bodies compounded of them; even so hard as never to wear or break in pieces."

Hooke was followed by Daniel Bernoulli,‡ who is often credited with many of the discoveries of Gassendi and Hooke. Bernoulli, again supposing that gas-pressure results from the impacts of particles on the boundary, was able to deduce Boyle's law for the relation between pressure and volume. In this investigation the particles were supposed to be infinitesimal in size, but Bernoulli further attempted to find a general relation between pressure and volume when the particles were of finite size, although still absolutely hard and spherical.

After Bernoulli, there is little to record for almost a century. Then we find Herapath§ (1821), Waterston‖ (1845), Joule¶ (1848), Kronig** (1856), Clausius (1857) and Maxwell (1859) taking up the subject in rapid succession.

* See § 50, below.

† *Opticks*, Query 31 (this did not appear until the second edition of the *Opticks*, 1718).

‡ Daniel Bernoulli, *Hydrodynamica*, Argentoria, 1738: Sectio decima, "De affectionibus atque motibus fluidorum elasticorum, praecipue autem aeris."

§ *Annals of Philosophy* (2), **1**, p. 273.

‖ *Phil. Trans. Roy. Soc.* **183** (1892), p. 1. Waterston presented a long paper to the Royal Society in 1845, but this contained many inaccuracies and so was not published until Lord Rayleigh secured its publication, for what was then a purely historical interest, in 1892.

¶ *British Association Report*, 1848, Part II, p. 21; *Memoirs of the Manchester Literary and Philosophical Society* (2), **9**, p. 107.

** *Poggendorff's Annalen.*

In his first paper* Clausius calculated accurately the relation between the temperature, pressure and volume in a gas with molecules of infinitesimal size; he also calculated the ratio of the two specific heats of a gas in which the molecules had no energy except that of their motion through space. In 1859, Clerk Maxwell read a paper before the British Association at Aberdeen,† in which the famous Maxwellian law of distribution of velocities made its first appearance, although the proof by which Maxwell attempted to establish it is now universally agreed to have been invalid.‡ In the hands of Clausius and Maxwell the theory developed with great rapidity, so that to write its history from this time on would be hardly less than to give an account of the subject in its present form.

The Three States of Matter

2. Most substances are capable of existing in three distinct states, which we describe as solid, liquid and gaseous. The typical example is water, with its three states of ice, water and steam. It is natural to conjecture that the three states of matter correspond to three different types or intensities of motion of the fundamental particles of which the matter is composed, and it is not difficult to see how the necessity for these three different states may arise.

We know that two bodies cannot occupy the same space; if we try to make them do so, repulsive forces come into play, and keep the two bodies apart. If matter consists of innumerable particles, these forces must be the aggregate of the forces from individual particles. These particles can, then, exert forces on one another, and the forces are repulsive when the particles are pushed sufficiently close to one another. If we try to tear a solid body into pieces, another set of forces comes into play—the forces of cohesion. These also indicate the existence of forces between individual particles, but the force between two particles is no longer one of repulsion; it is now one of attraction. The fact that a solid body, when in its natural state, resists both compression

* "Ueber die Art der Bewegung welche wir Wärme nennen", *Pogg. Ann.* **100**, p. 353.

† *Phil. Mag.* Jan. and July 1860; *Collected Works*, **1**, p. 377.

‡ See Appendix I, p. 296.

and dilatation, shews that the force between particles changes from one of repulsion at small distances to one of attraction at greater distances, as was pointed out by Boscovitch in 1763.*

3. *The Solid State.* Somewhere between these two positions there must be a position of stable equilibrium in which two particles can rest in proximity without either attracting or repelling one another. If we imagine a great number of particles placed in such proximity, and so at rest in their positions of equilibrium, we have the kinetic theory conception of a mass of matter in the solid state—a solid body. Modern X-ray technique makes it possible to study both the nature and arrangement of these particles. In solid bodies of crystalline structure, the "particles" are atoms and electrons, arranged in a regular three-dimensional pattern;† in conductors the electrons are free to thread their way between the atoms.

When the particles which constitute a solid body oscillate about their various positions of equilibrium, we say that the body possesses heat. The energy of these oscillatory motions is, in fact, the heat-energy of the body. As the oscillations become more vigorous, we say that the temperature of the body increases.

We may make a definite picture by supposing that the oscillatory motions are first set up by rubbing two solid objects together. We place the surfaces of the two bodies so close to one another that the particles near the surface of one exert perceptible forces on the particles near the surface of the other; we then move the surfaces over one another, so that the forces just mentioned draw or push the surface particles from their positions of equilibrium. At first, the only particles to be disturbed will be those which are in the immediate neighbourhood of the parts actually rubbed, but gradually the motion of these parts will induce motion in the adjoining regions, until ultimately the motion spreads over the whole mass. This motion represents heat which was, in the first instance, generated by friction, and then spread by conduction through the whole mass.

* Theoria Philosophiae Naturalis (Venice, 1763; English translation, Open Court, Chicago and London, 1922). See especially §§ 74 ff.

† See W. L. Bragg, *Crystal Structure*, and innumerable other books and papers.

As a second example, we may imagine that two solid bodies, both devoid of internal motion, impinge one upon the other—as for instance a hammer upon a clock-bell. The first effect of this impact will be that trains of waves are set up in the two bodies, but after a sufficient time the wave character of the motion will become obliterated. Motion of some kind must, however, persist in order to account for the energy of the original motion. This original motion will, in actual fact, be replaced by small vibratory motions in which the particles oscillate about their positions of equilibrium—according to the kinetic theory, by heat motion. In this way the kinetic energy of the original motion of the solid bodies is transformed into heat-energy.

4. *The Liquid State.* If a solid body acquires more heat, the energy of its vibrations will increase, so that the excursions of its particles from their positions of equilibrium will become larger. If the body goes on acquiring more and more heat, some of the particles will ultimately be endowed with so much kinetic energy that the forces from the other particles will no longer be able to hold them in position; they will then, to borrow an astronomical term, escape from their orbits, and move to other positions. When a considerable number of particles are doing this, the application of even a small force, provided it is continued for a sufficient length of time, can cause the mass to change its shape; it does this by taking advantage time after time, as opportunity occurs, of the weakness of the forces tending to retain individual particles.

If still more heat is provided, a greater and greater number of the particles will move freely about; finally, when all the particles are all doing this, the body has attained the state we describe as liquid.

So long as the body is in the solid state, the particles which execute vibrations will usually be either isolated electrons or atoms. In the liquid state it is comparatively rare for either electrons or atoms to move as independent particles, because the forces binding these into molecules are usually too strong to be overcome by the heat-motion; thus the particles which move independently in a liquid are generally complete molecules.

Until recently it was supposed that the molecules of a liquid moved at random, and shewed a complete disorder in their arrangement—in sharp contrast to the orderly arrangement of the particles in a solid. Recent investigations suggest the need for modifying this view. A molecule in a liquid is probably acted on all the time by about as many other molecules as it would be if in the solid state. The forces from these neighbouring molecules imprison it in a cell from which it only rarely escapes. The main difference between a solid and a liquid is not that between captivity and freedom; it is only that the particle of a solid is held in fetters *all* the time, while that of a liquid is held in fetters *nearly all* the time, living a life of comparative freedom only in the very brief intervals between one term of imprisonment and the next. Further X-ray technique has shewn that there is a certain degree of regularity and order in the arrangement of the liquid molecules in space.*

This knowledge is derived only from a statistical study of the molecules of a liquid; no known technique makes it possible to see the wanderings of individual molecules. Perhaps this is not surprising, since even the largest of molecules are beyond the limits of vision in the most powerful of microscopes. But if a number of very small solid particles—as, for instance, of gamboge or lycopodium—are placed in suspension in a liquid, these particles are set into motion as the moving molecules of the liquid collide with them and hit them about, now in this direction and now in that. These latter motions can be seen through a microscope, so that the solid particles act as indicators of the motions of the molecules of the liquid, and so give a very convincing, even if indirect, proof of the truth of the kinetic theory conception of the liquid state. They are called Brownian movements, after the English botanist, Robert Brown.

For, in 1828 Brown had suspended grains of pollen in water, and examined the mixture through a microscope. He found that the pollen grains were engaged in an agitated dance, which was to all appearances continuous and interminable. His first thought was that he had found evidence of some vital property in the pollen,

---

* See in particular, "Recent Theories of the Liquid State", N. F. Mott and R. W. Gurney, *Physical Society Reports on Progress in Physics* (C.U. Press), **5** (1939), p. 46.

but he soon found that any small particles, no matter how non-vital, executed similar dances. The true explanation, that the particles were acting merely as indicators of the molecular motion of the liquid in which they were immersed, was given by Delsaux in 1877 and again by Gouy in 1888. A full mathematical theory of the movements was developed by Einstein and von Smoluchowski about 1905.*

In 1909 Perrin suspended particles of gamboge in a liquid of slightly lower density, and found that the heavy particles did not sink to the bottom of the lighter liquid; they were prevented from doing so by their own Brownian movements. If the liquid had been infinitely fine-grained, with molecules of infinitesimal size and weight, every solid particle would have had as many impacts from above as below; these impacts, coming in a continuous stream, would have just cancelled one another out, so that each particle would have been free to fall to the bottom under its own weight. But when they were bombarded by molecules of finite size and weight, the solid particles were hit, now in one direction and now in another, and so could not lie inertly on the bottom of the vessel. From the extent to which they failed to do this, Perrin was able to form an estimate of the weights of the molecules of the liquid (§ 16, below) and this agreed so well with other estimates that there could be but little doubt felt as to the truth either of the kinetic theory of liquids, or of the associated explanation of the Brownian movements.

A molecule of a liquid which has escaped from its orbit in the way described on p. 6 may happen to come near to the surface of the liquid, in which case it may escape altogether from the attraction of the other molecules, just as a projectile which is projected from the earth's surface with sufficient velocity may escape from the earth altogether. When this happens the molecule leaves the liquid, and the liquid must continually diminish both in mass and volume owing to the loss of such molecules, just as the earth's atmosphere continually diminishes owing to the escape of rapidly moving molecules from its outer surface. Here we have the kinetic theory interpretation of the process of evaporation, the vapour being, of course, formed by the escaped molecules.

* See § 180, below.

If the liquid is contained in a closed vessel, each escaping molecule must in time strike the side or top of the vessel; its path is now diverted, and it may fall back again into the liquid after a certain number of impacts. In time a state may be reached such that as many molecules fall back in this way as escape by evaporation; we now have, according to the kinetic theory, a liquid in equilibrium with its own vapour.

5. *The Gaseous State*. On the other hand, it is possible for the whole of the liquid to be transformed into vapour in this way, before a steady state is reached. Here we have the kinetic theory picture of a gas—a crowd of molecules, each moving on its own independent path, entirely uncontrolled by forces from the other molecules, although its path may be abruptly altered as regards both speed and direction, whenever it collides with another molecule or strikes the boundary of the containing vessel. The molecules move so swiftly that even gravity has practically no controlling effect on their motions. An average molecule of ordinary air moves at about 500 metres a second, so that the parabola which it describes under gravity has a radius of curvature of about 25 kilometres at its vertex, and even more elsewhere. This is so large in comparison with the dimensions of any containing vessel that we may, without appreciable error, think of the molecules as moving in straight lines at uniform speeds, except when they encounter either other molecules or the walls of the containing vessel. This view of the nature of a gas explains why a gas spreads immediately throughout any empty space in which it is placed; there is no need to suppose, as was at one time done, that this expansive property is evidence of repulsive forces between the molecules (cf. § 50, below).

As with a liquid, so with a gas, there is no absolutely direct evidence of the motions of individual molecules, but an indirect proof, at one remove only, is again provided by the Brownian movements. For these occur in gases as well as in liquids; minute particles of smoke* and even tiny drops of oil floating in a gas may be seen to be hit about by the impact of the molecules of the gas.

---

* Andrade and Parker, *Proc. Roy. Soc.* A, 159 (1937), p. 507.

Recently E. Kappler* constructed a highly sensitive torsion-balance, in which the swinging arm was free to oscillate in an almost perfect vacuum. The swinging arm had a moment of momentum of 0·235 millionth of a gm. cm.², and when it swung in a gas at pressure of only a few hundreds of a millimetre of mercury, its period of oscillation was about 15 seconds. When the oscillations were recorded on a moving photographic film, it was seen that they did not proceed with perfect regularity. The speed of motion of the arm experienced abrupt changes, and, as we shall see later (p. 129), there can be no doubt that these were caused by the impacts of single molecules of the gas.

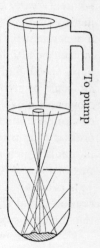

Perhaps, however, the simplest evidence of the fundamental accuracy of the kinetic theory conception of the gaseous state is to be found in experiments of a type first performed by Dunoyer.† He divided a cylindrical tube into three compartments by means of two partitions perpendicular to the axis of the tube, these partitions being pierced in their centres by small holes, as in fig. 1. The tube was fixed vertically, and all the air pumped out. A small piece of sodium was then introduced into the lowest compartment, and heated to a sufficient temperature to vaporise it. Molecules of sodium are now shot off, and move in all directions.

Fig. 1

Most of them strike the walls of the lowest compartment of the tube and form a deposit there, but a few escape through the hole in the first partition, and travel through the second compartment of the tube. These molecules do not collide with one another, since their paths all radiate from a point—the small hole through which they have entered the compartment; they travel like rays of light issuing from a source at this point. Some travel into the upper-most compartment through the opening in the second partition, and when they strike the top of the tube, make a deposit there.

* *Ann. d. Phys.* **31** (1938), p. 377.

† L. Dunoyer, *Comptes Rendus*, **152** (1911), p. 592, and *Le Radium*, **8** (1911), p. 142.

This is found to be an exact projection of the hole through which they have come. The experiment may be varied by removing the upper partition and suspending a small object in its place. This is found to form a "shadow" on the upper surface of the tube; if the hole in the first diaphragm is of appreciable size, an umbra and penumbra may even be discernible.

Stern* and Gerlach as well as a band of collaborators at Hamburg, and many workers elsewhere, have expanded this simple experiment into the elaborate technique known as "molecular rays". The lower compartment of Dunoyer's tube is replaced by an electric oven or furnace in which a solid substance can be partly transformed into vapour (see fig. 2). Some of the molecules of this vapour pass through a narrow slit in a diaphragm above the oven, and, as before, a small fraction of these pass later through a second parallel slit in a second diaphragm. These form a narrow beam of molecules all moving in the same plane (namely the plane through the two parallel slits) and

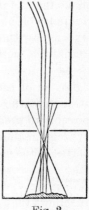

Fig. 2

this beam can be experimented on in various ways, some of which are described later in the present book.

Mechanical Illustration of the Kinetic Theory of Gases

6. It is important to form as clear an idea as possible of the conception of the gaseous state on which the kinetic theory is based, and this can best be done in terms of simple mechanical illustrations.

We still know very little as to the structure or shape of actual molecules, except that they are complicated structures of electrons and protons, obeying laws which are still only imperfectly unravelled. At collisions they obey the laws of conservation of momentum and energy, at least to a very close approximation. They also behave like bodies of perfect elasticity. For, as we shall shortly see, the pressure of a gas gives a measure of the

---

* *Zeitschr. f. Phys.* **39** (1926), p. 751.

total kinetic energy of motion of its molecules, and this is known to stay unaltered through millions of millions of collisions.

Since, however, it is desirable to have as concrete a representation as possible before the mind, at least at the outset, we shall follow a procedure which is very usual in the development of the kinetic theory, and agree for the present to picture a molecule as a spherical body of unlimited elasticity and rigidity—to make the picture quite definite, let us say a steel ball-bearing or a minute billiard ball. The justification for this procedure lies in its success; theory predicts, and experiment confirms, that an actual gas with highly complex molecules behaves, in many respects, like a much simpler imaginary gas in which the molecules are of the type just described. Indeed, one of the most striking features of the kinetic theory is the extent to which it is possible to predict the behaviour of a gas as a whole, while remaining in almost complete ignorance of the behaviour and properties of the molecules of which it is composed; the reason is that many of the properties of gases depend only on general dynamical principles, such as the conservation of energy and momentum. It follows that many of the results of the theory are true for all kinds of molecules, and so would still be true even if the molecules actually were billiard balls of infinitesimal size.

Since it is easier to think in two dimensions than in three, let us begin with a representation of molecular motions in two dimensions. As the molecules of the gas are to be represented by billiard balls, let us represent the vessel in which the gas is contained by a large billiard table. The walls of the vessel will of course be represented by the cushions of the table, and if the vessel is closed, the table must have no pockets. Finally, the materials of the table must be of such ideal quality that a ball once set in motion will collide an indefinite number of times with the cushions before being brought to rest by friction and other passive forces. A great number of the properties of gases can be illustrated with this imaginary apparatus.

If a large number of balls are taken and started from random positions on the table with random velocities, the resulting state of motion will give a representation of what is supposed to be the condition of matter in its gaseous state. Each ball will be continually colliding with the other balls, as well as with the cushions of the table. The velocities of the balls will be of the most varying kinds; a ball may be brought absolutely to rest at one instant, while at another, as the result of a succession of favourable collisions, it may possess a velocity far in excess of the average. We shall, in due course, investigate how the velocities of such a collection of balls are distributed about the mean velocity, and shall find that, no matter how the velocities are arranged at the outset, they will tend, after a sufficient number of collisions, to group themselves according to a law which is very similar to the well-known law of trial and error—the law which governs the grouping in position of shots fired at a target.

If the cushions of the table were not fixed in position, the continued impacts of the balls would of course drive them back. Clearly, then, the force which the colliding balls exert on the cushions represents the pressure exerted by the gas on the walls of the containing vessel. Let us imagine a movable barrier placed initially against one of the cushions, and capable of being moved parallel to this cushion. Moving this barrier forward is equivalent to decreasing the volume of the gas. If the barrier is moved forwards while the motion of the billiard balls is in progress, the impacts both on the movable barrier and on the three fixed cushions will become more frequent—because the balls have a shorter distance to travel, on the average, between successive impacts. Here we have a mechanical representation of the increase of pressure which accompanies a diminution of volume in a gas. To take a definite instance, let us suppose that the movable barrier is moved half-way up the table. As the space occupied by the moving balls is halved, the balls are distributed twice as densely as before in the restricted space now available to them. Impacts will now be twice as frequent as before, so that the pressure on the barrier is doubled. Halving the space occupied by our quasi-gas has doubled the pressure—an illustration of Boyle's law that the pressure varies inversely as the volume. This, however, is only true if the billiard balls are of infinitesimal size; if they are of appreciable size, halving the space accessible to them may more than double the pressure, since the extent to which they get in one another's way is more than doubled. At a later stage, we shall have to discuss how far the theoretical law connecting the pressure and density in a gas constituted of molecules of finite size is in agreement with that found by experiment for an actual gas.

Let us imagine the speed of each ball suddenly to be doubled. The motion proceeds exactly as before,* but at double speed—just as though we ran a cinematograph film through a projector at double speed. At each point of the boundary there will be twice as many impacts per second as before, and the force of each impact will be twice as great. Thus the pressure at the point will be four times as great. If we increase the speed of the balls in any other ratio, the pressure will always be proportional to the square of the speed, and so to the total kinetic energy of the balls. This illustrates the general law, which will be proved in due course, that the pressure of a gas is proportional to the kinetic energy of motion of its molecules. As we know, from the law of Charles, that the pressure is also proportional to the absolute temperature of the gas, we see that the absolute temperature of a gas must be measured by the kinetic energy of motion of its molecules—a result which will be strictly proved in due course (§ 12).

Still supposing the barrier on our billiard table to be placed half-way up the table, let us imagine that the part of the table which is in front of the barrier is occupied by white balls moving with high speeds, while the part behind it is similarly occupied by red balls moving with much smaller speeds. This corresponds to dividing a vessel into two separate chambers, and filling one with a gas of one kind at a high temperature, and the other with a gas of a different kind at a lower temperature. Now let the barrier across the billiard table suddenly be removed. Not only will the white balls immediately invade the part which was formerly occupied only by red balls, and vice versa, but also the rapidly moving white balls will be continually losing energy by collision with the slower red balls, while the red of course gain energy through impact with the white. After the motion has been in progress for a sufficient time, the white and red balls will be

---

* In actual billiards the angles of course depend on the speed, but this is a consequence of the imperfect elasticity of actual balls. With ideal balls of perfect elasticity, the course of events would be as stated above.

equally distributed over the whole of the table, and the average velocities of the balls of the two colours will be the same. Here we have simple illustrations of the diffusion of gases, and of equalisation of temperature. The actual events occurring in nature are, however, obviously more complex than those suggested by this analogy, for molecules of different gases differ by something more fundamental than mere colour.

Next let the barrier be replaced by one with a sloping face, so that a ball which strikes against it with sufficient speed will run up this face and over the top. Such a ball will be imprisoned on the other side, but one which strikes the slanting face at slower speed may run only part way up, and then come down again and move as though it had been reflected from the barrier. The barrier is in effect a mechanism for sorting out the quick-moving from the slow-moving balls.

Let the space between the sloping barrier and the opposite cushion of the table be filled with balls moving at all possible speeds to represent the molecules of a liquid. A ball which runs against the sloping face of the barrier with sufficient speed to get over the top will represent a molecule which is lost to the liquid by evaporation. The total mass of the liquid is continually diminishing in this way; at the same time, since only the quick-moving molecules escape, the average kinetic energy of the molecules which remain in the liquid is continually being lowered. Thus, as some molecules of the liquid evaporate, the remainder of the liquid decreases in temperature. Here we have a simple illustration of the cooling which accompanies evaporation—a process which explains the action of nearly all modern refrigerating machinery.

7. One further question must be considered. No matter how elastic the billiard balls and table may be, the motion cannot continue indefinitely. In time its energy will be frittered away, partly perhaps by frictional forces such as air-resistance, and partly by the vibrations set up in the balls by collisions. The energy dissipated by air-resistance becomes transformed into energy in the air; the energy dissipated by collision is transformed into energy of internal vibrations of the billiard balls. What, then, does this represent in the gas, and how is it that a gas, if consti-

tuted as we have supposed, does not in a very short time lose the energy of translational motion of its molecules, and replace it by energy of internal vibrations of these molecules, and of radiation travelling through the surrounding space?

The difficulties raised by this and similar questions formed a most serious hindrance to the progress of the kinetic theory for many years. Maxwell drew attention to them, and Kelvin, Rayleigh and many others worried over them, but it was only after the introduction of the quantum theory by Planck and his followers in the early years of the present century, that it became possible to give anything like a satisfactory explanation. This explanation is, in brief, that the analogy between billiard balls and molecules fails as soon as we begin to consider questions of internal vibrations and the transfer of their energy to the surrounding space. The analogy, which has served us well for a long time, breaks down at last. For the motion of billiard balls, as of all objects on this scale of size, is governed by the well-known Newtonian laws, whereas the internal motions of molecules, and their transfer of energy to the surrounding space in the form of radiation, are now believed to be governed by an entirely different system of laws. In the present book, we shall develop the kinetic theory as far as it can be developed without departing from the Newtonian laws, and shall notice the inadequacy of these laws in certain problems. The newer system of laws, constituting the modern theory of quantum mechanics, is beyond the scope of the present book.

Chapter II

A PRELIMINARY SURVEY

The Pressure in a Gas

8. The present chapter will contain a preliminary discussion of some of the principal problems of the kinetic theory. We shall not go into great detail, or strive after great mathematical accuracy or rigour—this is reserved for the more intensive discussions of later chapters. Our immediate object is merely to make the reader familiar with the main concepts of the theory, and to give him a rapid bird's-eye view of the subject, so as to ensure his not losing his bearings in the more complete discussions which come later.

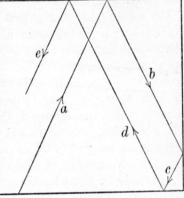

Fig. 3

The main actors in the drama of the kinetic theory are rapidly-moving molecules; the principal events in their lives are collisions with one another and with the boundary of a containing vessel. Let us consider the latter events first.

To begin with the very simplest conditions, let us first imagine that a single molecule is moving inside a cubical containing vessel of edge l, and colliding with the six faces time after time. We shall suppose that the molecule is infinitesimal in size, and also perfectly elastic, so that no energy will be lost on collision with the boundary. Thus the velocity of motion, which we shall denote by c, will retain the same value throughout the whole motion. We shall also suppose that at each collision the molecule bounces off the wall of the vessel at the same angle as that at which it struck. Thus we get a succession of paths such as a, b, c, d, e, ... in fig. 3.

We can find the lengths of successive paths by a very simple artifice. Let fig. 4 represent a honeycomb of cells, each of which is a cube of edge l. Draw the first path a of the molecule in the first cell, exactly as it is described in the actual cubical vessel, and extend the line a indefinitely through the honeycomb of cells. Then it is easy to see that the intercepts $b, c, d, e, \ldots$ made by successive cells will be exactly equal to the paths $b, c, d, e, \ldots$ of the actual molecule in the real vessel. If we imagine a point starting along the path a at the instant at which the actual molecule starts in the closed vessel, and moving at the same speed c as the real molecule, then the paths $a, b, c, d, e, \ldots$ in the honeycomb will be described at exactly the same time as the paths $a, b, c, d, e, \ldots$ in the real vessel.

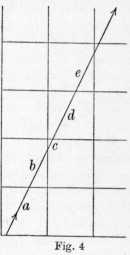

Fig. 4

Let rectangular coordinates Ox, Oy, Oz be taken in the honeycomb, parallel to the three directions of the edges of the cube, and let λ, μ, ν be the direction cosines of the line $abcde\ldots$. In the course of unit time, we must suppose our moving point to describe a distance c along this line, so that it will travel a distance $c\lambda$ parallel to Ox, $c\mu$ parallel to Oy and $c\nu$ parallel to Oz. In travelling the distance $c\lambda$ parallel to Ox, it will encounter faces of the honeycomb perpendicular to Ox at equal distances l apart. Each such encounter will of course coincide with a collision between the real molecule and a face perpendicular to Ox. The number of such collisions in unit time is therefore $c\lambda/l$.

At each such collision, the momentum parallel to Ox is reversed. If m is the mass of the molecule, this is of amount $mc\lambda$, so that the total transfer of momentum, or "impact", between the molecule and the face of the vessel is $2mc\lambda$. The total of all such impacts between the moving molecule and the two faces perpendicular to Ox must be

$$2mc\lambda \times \frac{c\lambda}{l} = \frac{2mc^2\lambda^2}{l}.$$

There are similar impacts with the other pairs of faces, of total amounts

$$\frac{2mc^2\mu^2}{l} \quad \text{and} \quad \frac{2mc^2\nu^2}{l}.$$

Thus the total of all the impacts exerted by the molecule on all the six faces in unit time is

$$\frac{2mc^2}{l}(\lambda^2+\mu^2+\nu^2) = \frac{2mc^2}{l}.$$

This, the total impact exerted in unit time, is also the total pressure exerted on the six faces of the cube.

If there are a very great number of molecules inside the vessel, of masses $m_1, m_2, m_3, \ldots$ moving with different velocities $c_1, c_2, c_3 \ldots$, and so small as never to collide with one another, the total impact they exert on all six faces is

$$\frac{2}{l}(m_1 c_1^2 + m_2 c_2^2 + m_3 c_3^2 + \ldots). \qquad \ldots\ldots(1)$$

If these molecules move in all directions at random, they will obviously exert equal pressures on the six faces of the cube. As the total area of the six faces is $6l^2$, the pressure p per unit area is given by

$$p = \frac{1}{3l^3}(m_1 c_1^2 + m_2 c_2^2 + \ldots).$$

The numerator in this fraction is twice the total kinetic energy of motion of all the molecules in the gas, while the denominator is three times the volume of the gas. Thus we have

$$\text{pressure} = \frac{2 \times \text{kinetic energy}}{3 \times \text{volume}}, \qquad \ldots\ldots(2)$$

and see that the *pressure is equal to two-thirds of the kinetic energy per unit volume.*

Since kinetic energies are additive, pressures must also be additive. Thus *the pressure exerted by a mixture of gases is the sum of the pressures exerted by the constituents of the mixture separately.* This is Dalton's law.

If the volume of the vessel is allowed to change, while the molecules and their energy of motion are kept the same, we see that *the pressure varies inversely as the volume.* This is Boyle's law.

If we write C^2 for the average value of c^2 throughout the gas, the average being taken by mass, we have

$$m_1 c_1^2 + m_2 c_2^2 + \ldots = (m_1 + m_2 + \ldots)\, C^2,$$

while

$$m_1 + m_2 + \ldots = \rho v,$$

where v is the volume and ρ the density of the gas. From equation (2) the pressure is now given by

$$p = \tfrac{1}{3}\rho C^2, \qquad\qquad\qquad \ldots\ldots(3)$$

so that

$$C^2 = \frac{3p}{\rho}. \qquad\qquad\qquad \ldots\ldots(4)$$

With this formula we can calculate the speed of molecular motion for a gas in any physical condition we please. Ordinary air at room temperature and at a pressure of one atmosphere is found to have a density of $0\cdot00123$ gramme per litre. Inserting these values of ρ and p in equation (4), we find that the molecular velocity is about 500 metres per second, which is roughly the speed of a rifle bullet.

9. We have not yet shewn that the pressure has the same value at all regions on the surface of the vessel, nor that its amount depends only on the volume of the vessel, independently of its shape—these facts are established in the next chapter (§ 33). Neither have we taken account of the collisions of the various molecules with one another, but it is easy to see that these do not affect the result. For a collision between two molecules does not of itself produce any pressure on the boundary, and neither does it affect the value of expression (2), since energy is conserved at a collision.

It may at first seem strange that the pressure should depend on the state of things throughout the whole of the gas in a vessel, and not only on the state of things close to the boundary on which the pressure is exerted. One molecule, for instance, in the far interior of the gas may be moving with immense velocity. How, it may be asked, can this molecule exert a correspondingly immense pressure on the boundary from which it is completely disconnected? The answer is, of course, that after a short time either the molecule will itself strike the boundary or else, through the medium of collisions, will communicate its high energy to other molecules which will strike the boundary in due course. For we must notice that formula (2) does not give the pressure at a single instant of time, but only the pressure averaged through one second, or of course through some other comparatively long interval of time.

A more serious question is whether we were justified in assuming that the molecules bounce off the walls of the containing vessel at the same angle, and with the same speed, at which they met this wall. The assumption was introduced solely for the sake of simplicity, and is probably not true (see § 34, below). But it is in no way essential to the proof. All that is necessary is that the molecules should, *on the average*, leave the walls with energy equal to that with which they arrive. The aggregate of the impacts which the molecules exert on the walls before they are brought to rest is then given by half of expression (1); the other half represents the impacts which the walls exert on the molecules to restore their motion. Then the total pressure is as before.

Even the supposition that the *average* energy of a molecule is unchanged by impact with the boundary may not of course be strictly true. If a very hot gas is put into a very cold vessel, we know that the vessel will acquire heat from the gas, and will continue to do so until the temperatures of the gas and vessel are equal. The transfer of heat from the gas to the vessel can only take place when molecules collide with the wall of the vessel, so that they do not, on the average, leave it with as much energy as they brought to it. This does not have any great influence on the calculation of the pressure, because the transfer of heat is so slow a process that very little energy can be transferred at a single collision. But it points the way to other questions of interest.

Equipartition of Energy

10. Let us consider in some detail the transfer of energy at the collisions of two molecules, which will be supposed to be smooth hard spheres of perfect elasticity, and will ultimately be taken to belong to the wall and the gas respectively. Let us take the line of impact at collision to be the axis Ox, this not necessarily coinciding with our former Ox, which was perpendicular to one of the walls.

Let the molecules have masses m, m', and let their components of velocity before collision be

$$u, v, w \text{ and } u', v', w'.$$

Since the impact occurs along the axis Ox, the components of velocity parallel to Oy and Oz are unaltered, so that we may suppose the velocities after the collision to be

$$\bar{u}, v, w \text{ and } \bar{u}', v', w'.$$

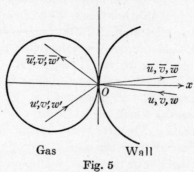

Fig. 5

The equations of energy and momentum now take the simple forms

$$\tfrac{1}{2}mu^2 + \tfrac{1}{2}m'u'^2 = \tfrac{1}{2}m\bar{u}^2 + \tfrac{1}{2}m'\bar{u}'^2,$$

$$mu + m'u' = m\bar{u} + m'\bar{u}',$$

or, by a slight transposition,

$$m(\bar{u}^2 - u^2) = -m'(\bar{u}'^2 - u'^2), \qquad \text{......(5)}$$

$$m(\bar{u} - u) = -m'(\bar{u}' - u'). \qquad \text{......(6)}$$

From these equations we obtain at once, by division of corresponding sides,

$$\bar{u} + u = \bar{u}' + u' \qquad \text{......(7)}$$

or

$$\bar{u}' - \bar{u} = -(u' - u),$$

shewing that the relative velocity is simply reversed at collision, a necessary consequence of perfect elasticity.

Solving equations (6) and (7), we find that the velocities after collision are given by

$$(m + m')\bar{u} = (m - m')u + 2m'u', \qquad \text{......(8)}$$

$$(m + m')\bar{u}' = (m' - m)u' + 2mu. \qquad \text{......(9)}$$

Now if m is the mass of the molecule of the wall, and m' that of the molecule of the gas, this collision results in the wall gaining energy of amount

$$\tfrac{1}{2}m(\overline{u}^2 - u^2) = \tfrac{1}{2}m(\overline{u} - u)(\overline{u} + u)$$

$$= \frac{2mm'}{(m+m')^2}(m'u' + mu)(u' - u)$$

$$= \frac{2mm'}{(m+m')^2}[(m'u'^2 - mu^2) + (m - m')uu'].$$

When a collision takes place, the velocity u of the wall molecule may be either positive or negative. Since this molecule has, on the whole, no continuous motion along the axis of x, and does not change its position except for small in-and-out oscillations, the average value of u must be zero. Thus if we average over a large number of collisions, the average value of uu' is zero, and we find as the mean gain of energy to the wall

$$\frac{2mm'}{(m+m')^2}(\overline{m'u'^2} - \overline{mu^2}),$$

where $\overline{mu^2}$ denotes the mean value of mu^2, and so on.

Thus the vessel gains in energy and so rises in temperature if the average value of $m'u'^2$ is greater than the average value of mu^2, and conversely. If the vessel and the gas have the same temperature, the vessel neither gains nor loses energy on the average, so that we have

$$\overline{mu^2} = \overline{m'u'^2}. \qquad\qquad \text{......(10)}$$

11. Suppose next that two kinds of gas are mixed in the vessel, the mass and velocity-components of the molecules of the second kind being denoted by m'', u'', v'', w'', and suppose further that both kinds of gas are at the same temperature as the vessel itself. Then there is no loss or gain of energy to the vessel through collisions with either kind of molecule, so that, in place of equation (10), we must have

$$\overline{m''u''^2} = \overline{mu^2} = \overline{m'u'^2}. \qquad\qquad \text{......(11)}$$

Since the molecules will be moving in all directions equally, we also have

$$\overline{m'v'^2} = \overline{m'w'^2} = \overline{m'u'^2}, \qquad\qquad \text{......(12)}$$

and there are similar equations for the second kind of molecules.

Equation (11) gives at once the relation

$$\overline{\tfrac{1}{2}m''(u''^2+v''^2+w''^2)} = \overline{\tfrac{1}{2}m'(u'^2+v'^2+w'^2)}. \quad \ldots\ldots(13)$$

Thus *when two gases are mixed at the same temperature, the average kinetic energy of their molecules is the same.* The less massive molecules must move, on the average, faster than the more massive, the difference of speed being such as to make the average kinetic energy the same in the two cases. This is a special case of a much wider theorem known as the equipartition of energy. The general theorem is of very wide application, depending as it does only on the general laws of dynamics. (See § 211 below.)

An illustration, of a rather extreme kind, is provided by the Brownian movements. Andrade and Parker* find that the particles of freshly-burned tobacco smoke have an average diameter of about $1\cdot6\times10^{-6}$ cm., so that their mass must be of the order of 10^{-18} gm., or 20,000 times the mass of a molecule of air. If the particles are suspended in air, their average speed will be about a 140th of that of the air molecules—say 3 metres a second. The combination of ultra-slow speed with ultra-large size makes it possible to study their motions in the microscope. Yet even here we do not see this actual motion, but only the motion resulting from a succession of free paths, each of which is performed with this velocity on the average.

Another illustration of the same dynamical principle can be found in astronomy, being provided by the motion of the stars in space. The stars move with very different speeds, but there is found to be a correlation between their speeds and masses, the lighter stars moving the faster, and this correlation is such that the average kinetic energy of stars of any specified mass is (with certain limitations) equal to that of the stars of any other mass. Here, as so often in astronomy, the stars may be treated as molecules of a gas, and the facts just stated seem to shew that the stars have been mixed long enough for all types of stars to have attained the same "temperature".

* *Proc. Roy. Soc.* A, **159** (1937), p. 515.

Temperature and Thermodynamics

12. Equation (13) has shewn that the average kinetic energy of motion of a molecule depends only on the temperature, and it is important to discover the exact law of this dependence.

We must first define temperature. Many ways of measuring temperature depend on the physical properties of particular substances. The ordinary thermometry, for instance, depends on the physical properties of water, glass and mercury. If we used alcohol or gas in our thermometer instead of mercury, we should obtain a slightly different temperature-scale. In fact, there are at least as many ways of measuring temperature as there are substances wherewith to measure it. But, outside all these, there is one more fundamental way which is independent of the properties of any particular substance. Clearly this is the way which should be used in the kinetic theory, since this is concerned with the properties of all substances equally.

According to thermodynamical theory, the addition of a small amount of heat-energy dQ to a mechanical system at a temperature θ produces an increase $dQ/f(\theta)$ in a physical quantity S, which is defined to be the "entropy" of the system. Here $f(\theta)$ is a function, so far unspecified, of the temperature θ, which may so far be measured in any way we please; a change in the method of measuring θ is compensated by a change in the form of the function f.

This expresses no property of nature or of matter so far. It merely defines S as the quantity obtained by successive algebraic additions of the quantity $dQ/f(\theta)$, and so as

$$S = \int \frac{dQ}{f(\theta)}.$$

But the investigations of Carnot* shewed that there is a certain particular form for $f(\theta)$, for which the quantity S defined in this way depends only on the physical state of the mechanical system concerned. If we start from any physical state A, and after a series of additions and subtractions of heat-energy bring the system back again to its original state A, then S will also

---

* *Réflexions sur la puissance motrice du feu* (English translation, Macmillan (1890)), or Preston, *Theory of Heat*, Chapter VIII, Section II.

return to its original value. Thus for such a cycle of changes, we must have

$$\int_{A}^{A} \frac{dQ}{f(\theta)} = 0.$$

This expresses Clausius' form of the second law of thermodynamics. The same thing may be expressed more mathematically by the statement that if

$$dS = \frac{dQ}{f(\theta)}$$

then dS is a "perfect differential". These relations are true for all mechanical systems, and so do not depend on the physical properties of any particular substances except in so far as these are involved in the measuring of θ.

Lord Kelvin pointed out in 1848* that the foregoing facts make it possible to build up a thermometer scale which shall be independent of the physical properties of all substances; we need only define the new temperature T by the equation

$$T = f(\theta).$$

The scale obtained in this way is known as the "absolute" scale. It is indefinite only to the extent of a multiplying constant, since T might equally well be chosen to be any multiple of $f(\theta)$. In practice, this constant is chosen so that the difference between the freezing and boiling points of water (under the same conditions as for the centigrade scale) is equal to 100 degrees. To this extent, and this only, the "absolute" scale is related to a particular substance. On this scale, the entropy S is defined by

$$dS = \frac{dQ}{T}.$$

We proceed to connect the results already obtained with this absolute scale of temperature.

Returning to our cubical vessel of volume l^3, let us suppose that one of its faces can slide in and out, like a piston in a cylinder, so that we can change the volume of the vessel. At first let it be of volume l^3, as before, and contain N molecules of average

* W. Thomson, *Proc. Camb. Phil. Soc.* (1848) and *Trans. Roy. Soc. Edinburgh* (1854).

kinetic energy $\bar{E}$, each molecule being a hard elastic sphere of infinitesimal size.

Now suppose that a small quantity of energy is added, in the form of heat, to the gas in the cylinder. Part of this may warm up the gas, so that the average energy of each molecule is increased from $\bar{E}$ to $\bar{E} + d\bar{E}$; as there are N molecules, this uses up an amount $N d\bar{E}$ of the energy. Another part of the added energy may be used in expanding the volume v of the gas. If the movable face increases its distance from the opposite face by dl, the volume will be increased from l^3 to $l^2(l + dl)$, an increase of volume dv equal to $l^2 dl$. As the gas exerts a pressure pl^2 on this movable face, there must also be a force pl^2 acting on the movable face from outside to hold it in position.

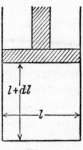

Fig. 6

When the gas expands to the extent just described, this force is pushed back a distance dl, and the energy consumed in doing this is $pl^2 dl$ or $p\,dv$. The sum of these two amounts of energy must be equal to the amount of heat-energy added to this gas. Calling this dQ, we must have

$$dQ = N d\bar{E} + p\,dv. \qquad \ldots\ldots(14)$$

On inserting the value already found for p in equation (2), this becomes

$$dQ = N d\bar{E} + \frac{2N\bar{E}}{3v}\,dv. \qquad \ldots\ldots(15)$$

Thus the value of dS is given by

$$dS = \frac{dQ}{T} = \frac{N}{T}\,d\bar{E} + \frac{2N\bar{E}}{3Tv}\,dv.$$

Since S depends only on $\bar{E}$ and v, we have

$$\frac{\partial S}{\partial \bar{E}} = \frac{N}{T}, \quad \frac{\partial S}{\partial v} = \frac{2N\bar{E}}{3Tv}.$$

We obtain the value of $\dfrac{\partial^2 S}{\partial \bar{E} \partial v}$ either by differentiating $\partial S/\partial v$ with respect to $\bar{E}$, or by differentiating $\partial S/\partial \bar{E}$ with respect to v. Equating the two values so obtained, we find

$$\frac{\partial}{\partial \bar{E}}\left(\frac{2N\bar{E}}{3Tv}\right) = 0.$$

This shews that $\bar{E}/T$ must be a constant, and gives us our first insight into the physical meaning of absolute temperature; we see that it is simply proportional to the average kinetic energy of motion of a molecule. We may in fact write

$$\frac{2}{3}\frac{\bar{E}}{T} = R, \qquad \ldots\ldots(16)$$

where R is a constant which is usually called Boltzmann's constant. We have already seen (§ 11) that $\bar{E}$ has the same value for all gases at any assigned temperature, so that R is the same for all gases; it is in fact a universal constant of nature. Its value, as we shall see later, is $1\cdot379 \times 10^{-16}$ in c.g.s. centigrade units.

We have thus found that the average kinetic energy of motion of each molecule is proportional to the absolute temperature, the exact relation being

$$\tfrac{1}{2}m(u^2 + v^2 + w^2) = \tfrac{3}{2}RT. \qquad \ldots\ldots(17)$$

Since the motions of the molecules can have no preference for particular directions in space, the mean values of u^2, v^2 and w^2 must all be equal, so that relation (17) may be expressed in the form

$$\overline{mu^2} = \overline{mv^2} = \overline{mw^2} = RT. \qquad \ldots\ldots(18)$$

The Gas-laws

13. With these relations we can at once express the relations between pressure, volume and temperature. In equation (2), namely

$$\text{pressure} = \frac{2 \times \text{kinetic energy}}{3 \times \text{volume}},$$

we may put the kinetic energy equal to $N\bar{E}$, where N is the total number of molecules and $\bar{E}$ is the average kinetic energy of each. The relation now becomes

$$p = \frac{2N\bar{E}}{3v} = \frac{NRT}{v}. \qquad \ldots\ldots(19)$$

If the N molecules consist of N_1 molecules of one kind, N_2 of a second kind, and so on, so that $N = N_1 + N_2 + \ldots$, we have

$$p = \frac{N_1RT}{v} + \frac{N_2RT}{v} + \frac{N_3RT}{v} + \ldots.$$

Thus the pressure exerted by a mixture of different gases is equal to the sum of the pressures which would be exerted by the constituents separately, which again brings us to Dalton's law (§ 8). We have already noticed that when the temperature is kept constant, p varies inversely as v (Boyle's law); equation (19) shews further that when the volume is kept constant, p varies as T, the absolute thermodynamic temperature (Charles' law). Further, when p is kept constant, v varies as T; if a mass of gas is kept at constant pressure, its volume will change in exact proportionality with the absolute temperature. Thus the absolute temperature is the temperature which would be read off a thermometer in which the expanding substance was a gas of the ideal kind we have been considering, the molecules being hard spheres of infinitesimal size, exerting no force except at collisions. This, indeed, expresses the fundamental principle of gas-thermometry. It enables us to fix the zero-point of the absolute temperature scale—the absolute zero of temperature. This is found to be

$$- 273 \cdot 2° \text{ C.} \qquad \qquad \text{......(20)}$$

At this temperature, $\bar{E} = 0$, which means that all the molecules are at rest; there is no molecular motion, so that the substance cannot be in the gaseous state. As the temperature is raised above this point, motion ensues, the average kinetic energy per molecule being always proportional to the temperature measured from this zero-point.

If we warm up the gas inside an enclosure which is kept at constant pressure, both p and v must remain constant, so that, from relation (19), NRT remains constant. It is interesting to notice that heating up the enclosure does not, as might at first be thought, increase the total quantity of heat inside it; as T increases N correspondingly decreases, so that all the heat we provide goes elsewhere. For instance, when we light a fire in a room, we do not increase the heat-content of the air of the room; this remains unchanged. By increasing the pressure, we drive some molecules out of the room, and the original amount of heat, now being distributed over a smaller number of molecules, gives more energy to each.

14. Formula (19), namely

$$pv = NRT,$$

shews that *the number of molecules in a gas of specified volume, temperature and pressure is the same for all gases, and so is independent of the nature of the gas.*

This is commonly known as Avogadro's law, having been proposed by Avogadro in 1811 as a hypothesis to explain the fact that gases unite, both by weight and by volume, in simple proportions which are measured by the ratios of small integral numbers.

Without the kinetic theory to help us, it would have been rather natural to think that, if a gas had specially massive molecules, the standard pressure of one atmosphere would be produced by a smaller number of molecules than in a gas with light molecules. The kinetic theory shews that at any specified temperature massive molecules move more slowly than light ones, so that in actual fact the pressure exerted per molecule is the same for all, namely RT/v. Thus the number of molecules per c.c. needed to produce one atmosphere pressure at the standard temperature of $0°$ C. is the same for all gases. This number, which is known as Loschmidt's number, will be denoted by N_0, and its numerical evaluation is naturally of great importance for the kinetic theory of matter.

Closely associated with it is another number, which measures the number of molecules in a gramme-molecule of any substance —i.e. in a mass of the substance in which the number of grammes is equal to the molecular weight of the substance. This is commonly known as Avogadro's number and will be denoted by N_1.

Evaluation of Avogadro's and Loschmidt's Numbers

15. These numbers can be evaluated in a great variety of ways. Virgo* has enumerated more than eighty different experimental determinations which had been made by 1933, and the number continually increases.

---

* *Science Progress*, **108** (1933), p. 634, or Loeb, *Kinetic Theory of Gases* (2nd edn, 1934), p. 408.

The numbers can be evaluated most accurately from electrolytic phenomena. When a mass of any substance is broken up electrolytically, a certain quantity of electricity is consumed, this being required to provide the electronic charges on the ions. Experiment shews that 9648·9 electromagnetic units of charge are required for every gramme-molecule of the substance. According to the most recent determinations, the electronic charge e—i.e. the charge on a single electron or ion—is given by

$$e = 4·803 \times 10^{-10} \text{ electrostatic units}$$
$$= 1·602 \times 10^{-20} \text{ electromagnetic units}$$

to an accuracy of within 1 part in 1,000.* Thus the above-mentioned 9648·9 electromagnetic units are equal to

$$N_1 = \frac{9648·9}{1·602 \times 10^{-20}} = 6·023 \times 10^{23} \qquad \ldots\ldots(21)$$

electronic units of charge. This, then, must be the number of molecules in a gramme-molecule of any substance, i.e. in a mass which contains a number of grammes equal to the molecular weight of the substance. Since the molecular weight of hydrogen is 2·016, a gramme-molecule of hydrogen weighs 2·016 grammes, so that the number of molecules of hydrogen to a gramme is

$$\frac{6·023 \times 10^{23}}{2·016} = 2·987 \times 10^{23}.$$

The density of hydrogen at standard temperature and pressure is 0·00008987, so that a cubic centimetre of hydrogen at standard temperature and pressure weighs 0·00008987 gramme. The number of molecules it contains is accordingly

$$N_0 = 0·00008987 \times 2·987 \times 10^{23} = 2·685 \times 10^{19}. \qquad \ldots\ldots(22)$$

This is Loschmidt's number.

16. Another, although far less exact, determination of N_1 can be made from observations on the Brownian movements in the way already explained. Perrin, who first developed the method, obtained values uniformly larger than those given above. Later

* Birge, *Nature*, **137** (1926), p. 187; *Phys. Rev.* 241 (1937), p. 241; H. R. Robinson, *Reports on Progress in Physics*, **4**, Physical Society (1938), p. 216; *Nature*, 23 July 1938, p. 159.

observations by H. Fletcher* gave the value $N_1 = 6.03 \times 10^{23}$, with a probable error of about 2 per cent of the whole, but a still more recent determination by Pospišil† gives $N_1 = 6.22 \times 10^{23}$. From experiments on the Brownian movements of his torsion-balance (§ 5), Kappler deduced as the value of Boltzmann's constant (see equation (60) below), $R = 1.36 \times 10^{-16}$. Combining this with the relation $pv = NRT$, we can deduce as the value of N_0,

$$N_0 = 2.67 \times 10^{19}.$$

Kappler estimated the probable error of his determination of R to be 3 per cent, so that there is a certain element of luck in the agreement between this and the true value of N_0.

17. The largeness of the numbers N_1 and N_0 gives a measure of the fine-grainedness of the molecular structure of matter. Some conception of the degree of fine-grainedness implied may perhaps be obtained from the following considerations.

The number of molecules in a drop of water one cubic millimetre in volume will be

$$\frac{N_1}{18.016 \times 10^3} = 3.34 \times 10^{19},$$

whence we can calculate that if the water is allowed to evaporate at such a rate that a million molecules leave it every second, the time required for the whole drop to evaporate will be 3.34×10^{13} seconds, which is more than a million years.

Again, a man is known to breathe out about 400 c.c. of air at each breath, so that a single breath of air must contain about 10^{22} molecules. The whole atmosphere of the earth consists of about 10^{44} molecules. Thus one molecule bears the same relation to a breath of air as the latter does to the whole atmosphere of the earth. If we assume that the last breath of, say, Julius Caesar has by now become thoroughly scattered through the atmosphere, then the chances are that each of us inhales one molecule of it with every breath we take. A man's lungs hold about 2000 c.c. of air, so that the chances are that in the lungs of each of us there are about five molecules from the last breath of Julius Caesar.

* *Phys. Rev.* **4** (1914), p. 440.
† *Ann. d. Phys.* **83** (1927), p. 735.

The $2\cdot685 \times 10^{19}$ molecules in a cubic centimetre of ordinary air move with an average speed of about 500 yards in a second. In a single second, then, the molecules in a cubic centimetre of air travel a total distance of 13×10^{21} yards, or 12×10^{18} kilometres, which is about 150,000 times the distance to Sirius.

18. Since a cubic centimetre of gas at 0° C. and at standard atmospheric pressure contains $2\cdot685 \times 10^{19}$ molecules, the average distance apart of adjacent molecules must be about

$$(2\cdot685 \times 10^{19})^{-\frac{1}{3}} \text{ cm.,}$$

or $3\cdot34 \times 10^{-7}$ cm. If we pass 100 km. up in the atmosphere—broadly speaking to auroral heights—we come to pressures of only about a millionth of an atmosphere; here the average distance of the molecules is only about 3×10^{-5} cm. Out in interstellar space the pressure is probably less by a further factor of about 10^{-15}; here adjacent molecules are, on the average, a few centimetres apart.

Molecular Masses

19. As there are $2\cdot987 \times 10^{23}$ molecules in a gramme of hydrogen, the mass of the hydrogen molecule must be $3\cdot347 \times 10^{-24}$ grammes, and that of the hydrogen atom $1\cdot673 \times 10^{-24}$ grammes.

The masses of other molecules will be in exact proportion to their molecular weights; that of oxygen, for instance, of molecular weight 32, is $53\cdot12 \times 10^{-24}$ grammes.

The Specific Heats of a Gas

20. As we have seen, the heat-energy of a gas is the aggregate of the energies of its separate molecules, and the temperature of the gas gives a measure of this energy. Thus to raise the temperature of a gas, the energy of its molecules must be increased. The amount of heat or of energy needed to raise the temperature of unit mass by 1° is called the specific heat of the gas; it may be measured either in heat- or in energy-units.

Equation (14) already obtained, namely

$$dQ = Nd\bar{E} + pdv, \qquad \ldots\ldots(23)$$

tells us how much energy is required to produce any specified change in the temperature and volume of a gas, and so provides a basis for a discussion of the specific heats of a gas.

Specific heat measurements are usually made under one of two alternative conditions:

(1) At constant volume, i.e. in a closed vessel.

(2) At constant pressure, i.e. in a vessel which is open to the air or other space at a fixed pressure.

To study specific heats measured at constant volume, we put $dv = 0$, and the last term in equation (23) is immediately obliterated; all the heat goes to increasing the energy of the molecules, and the equation becomes

$$dQ = N\,d\bar{E}.$$

The specific heat at constant volume, C_v, which is the heat required to raise unit mass of the gas through a temperature change of $1°$ C., is given by

$$C_v = \frac{dQ}{dT},$$

or again by
$$C_v = N\,\frac{d\bar{E}}{dT},$$

where N is now the number of molecules in a unit mass of the gas. Since $Nm = 1$, we may replace N by $1/m$. We have so far supposed the heat measured in energy-units. One calorie or heat-unit is equal to J energy-units, where J is the mechanical equivalent of heat, given by

$$J = 4{\cdot}184 \times 10^7,$$

so that if C_v is measured in heat-units, its value is

$$C_v = \frac{1}{Jm}\,\frac{d\bar{E}}{dT}. \qquad \ldots\ldots(24)$$

If the measurement of specific heat is made at constant pressure, the gas expands as its temperature is raised, and dv must no longer be put equal to zero. The value of the pressure is given by

$$pv = NRT,$$

so that when the pressure is kept constant, we obtain by differentiation

$$p\,dv = NR\,dT,$$

and equation (20) assumes the form

$$dQ = N\,d\bar{E} + NR\,dT,$$

whence the specific heat at constant pressure, C_p, measured in heat-units, is given by

$$C_p = \frac{1}{J}\frac{dQ}{dT} = \frac{1}{Jm}\left(\frac{d\bar{E}}{dT} + R\right). \qquad \ldots\ldots(25)$$

The difference and ratio (γ) of the specific heats are seen to be given by

$$C_p - C_v = \frac{R}{Jm}, \qquad \ldots\ldots(26)$$

$$\gamma = \frac{C_p}{C_v} = 1 + \frac{R}{d\bar{E}/dT}. \qquad \ldots\ldots(27)$$

If the molecules are hard elastic spheres, the value of $\bar{E}$ is $\frac{3}{2}RT$, and the value of C_p/C_v becomes $1\frac{2}{3}$.

Relation (26) is obeyed by most gases throughout a substantial range of physical conditions (Carnot's law). Also the ratio of the specific heats is equal to about $1\frac{2}{3}$ for the monatomic gases mercury, helium, argon, etc., suggesting that the atoms of these substances behave, in some respects at least, like the hard spherical balls which we have taken as our molecular model. On the other hand air and the permanent diatomic gases have a value of C_p/C_v equal to about $1\frac{2}{5}$ under normal conditions, suggesting that their molecules are not adequately represented by the hard spherical balls of our model. This is perhaps hardly surprising, seeing that these molecules are known to consist of two distinct and separable atoms.

21. Complex molecules of this kind may rotate or have internal motion of their parts, so that clearly we must suppose they can possess energy other than the kinetic energy of their motion; let us suppose that on the average this energy is β times the average kinetic energy of motion. The average total energy of a molecule $\bar{E}$ is now given by

$$\bar{E} = \tfrac{3}{2}RT(1 + \beta).$$

Carrying through the analysis as before, we find

$$C_v = \frac{3R}{2Jm}(1 + \beta), \qquad (28)$$

$$C_p = \frac{3R}{2Jm}(1+\beta)+\frac{R}{Jm}, \qquad \ldots\ldots(29)$$

in which β has been supposed not to vary with T.

The relation $\qquad C_p - C_v = \dfrac{R}{Jm},$

which is true for most gases, still holds, but in place of relation (27) we have

$$\gamma = \frac{C_p}{C_v} = 1 + \frac{2}{3(1+\beta)}. \qquad \ldots\ldots(30)$$

Thus the value of β can be deduced from the observed value of C_p/C_v for any gas. This gives us some knowledge of the structure of the molecules of gases, which will be produced and discussed in its proper place.

Maxwell's Law

22. So far we have been concerned only with the *average* kinetic energy, and the *average* speeds of molecules; we have had no occasion to think of the energies or speeds of individual molecules. And it is obvious that the speeds of motion of the various molecules cannot be all equal; even if they started equal, a few collisions would soon abolish their equality. For the solution of many problems, it is necessary to know how the velocities of motion are arranged round the mean, after so many collisions have occurred that the gas has reached its final steady state in which the distribution of velocities is no longer changed by collisions.

To study this problem, let us imagine that the molecules of a gas still move in a closed vessel, but that they are under the influence of some permanent field of force—gravitation will serve to fix a concrete picture in our minds, although it is more instructive to imagine something more general. Under these conditions, the velocity of a molecule will change continuously as it moves from point to point under the forces of gravity or the other permanent field of force, and will also change discontinuously each time it collides with another molecule.

Let us now fix our attention on a small group of molecules which occupies a small volume of space $dx\,dy\,dz$, defined by the condition that the x-coordinate lies within the small range from x to $x+dx$, while the y- and z-coordinates lie within similar ranges from y to $y+dy$, and from z to $z+dz$. We shall not concern ourselves with all the molecules inside the small element of volume, but only with those particular molecules of which the velocity components are approximately u, v, w; to be precise we shall confine our attention to molecules with velocities such that the u-component lies between u and $u+du$, while the v- and w-components lie within similar small ranges from v to $v+dv$ and w to $w+dw$. The number of molecules in this group will of course be proportional not only to the volume $dx\,dy\,dz$, but also to the product of the small ranges $du\,dv\,dw$. Thus it is proportional to the product of differentials

$$du\,dv\,dw\,dx\,dy\,dz.$$

It is also proportional to another factor of the nature of a "density", which specifies the number of molecules *per unit range* lying within the range in question. This factor naturally depends on the particular values of x, y, z and of u, v, w. Let us suppose that, whatever these values are, it is

$$f(u,v,w,x,y,z),$$

so that the number of molecules in the group under consideration is

$$f(u,v,w,x,y,z)\,du\,dv\,dw\,dx\,dy\,dz. \qquad \ldots\ldots(31)$$

As the motion of the gas proceeds, changes will occur in the values of x,y,z and u,v,w for individual molecules, and also in the values of dx,dy,dz,du,dv and dw for the group as a whole. Let us suppose that after a small time dt, these quantities have become

$$x',y',z',u',v',w' \text{ and } dx',dy',dz',du',dv',dw',$$

respectively. These are of course given by

$$x' = x+u\,dt, \text{ etc.} \qquad \ldots\ldots(32)$$

and
$$u' = u+X\,dt, \text{ etc.,} \qquad \ldots\ldots(33)$$

where X, Y, Z are the components of force per unit mass on a molecule at x, y, z.

If the gas is in a condition of steady motion, this group of molecules will exactly step into places which have just been vacated by a second group of molecules, which occupied these places at the beginning of the interval of time dt, and so will be equal in number to this second group. The number in the second group is, as in formula (31),

$$f(u', v', w', x', y', z') \, du' \, dv' \, dw' \, dx' \, dy' \, dz', \qquad \ldots\ldots(34)$$

so that if the gas is in a state of steady motion, expressions (34) and (31) must be equal.

The relation between u, v, w, x, y, z and u', v', w', x', y', z' is that expressed in equations (32) and (33). From these relations, it is fairly easy to shew that the product of the six differentials $du' \, dv' \, dw' \, dx' \, dy' \, dz'$ is precisely equal to the product $du \, dv \, dw \, dx \, dy \, dz$; we need not spend time over a detailed proof, since this would only constitute a very special case of a general theorem to be proved later (§ 206). The products of differentials in formulae (31) and (34) being equal, we must have the equation

$$f(u, v, w, x, y, z) = f(u', v', w', x', y', z'). \qquad \ldots\ldots(35)$$

Thus $f(u, v, w, x, y, z)$ is a quantity depending only on the values of u, v, w and of x, y, z, which does not change in value as a molecule follows out its natural motion without any collisions taking place. One such quantity is of course the energy of the molecule, which we may denote by E; clearly then

$$f(u, v, w, x, y, z) = E \qquad \ldots\ldots(36)$$

is a solution of equation (35). A more general solution is

$$f(u, v, w, x, y, z) = \Phi(E), \qquad \ldots\ldots(37)$$

where $\Phi(E)$ is any function whatever of E, as for instance its square or its logarithm. Here E, the total energy of a moving molecule, is given by

$$E = \tfrac{1}{2}m(u^2 + v^2 + w^2) + \chi, \qquad \ldots\ldots(38)$$

where χ is the potential energy of a molecule in the field of force, so that the forces acting on the molecule are given by

$$mX = \frac{\partial \chi}{\partial x}, \quad mY = \frac{\partial \chi}{\partial y}, \quad mZ = \frac{\partial \chi}{\partial z}.$$

It is quite easy to shew* that this expresses the most general solution possible of equation (35), but we need not delay over this proof, as a complete discussion of the whole problem will be given later (Chap. x).

23. We have now found that if the law of distribution is given by equation (37), where Φ represents any function whatever, the distribution of velocities will not be altered by the natural motion of the gas, so long as no collisions take place.

In general, however, the law of distribution will be altered by collisions, and the question arises whether there is any special form for the function Φ such that collisions have no effect. If so, on inserting this form of Φ in equation (37), we shall obtain a law of distribution which remains unaltered both by the natural motion of the gas and by the occurrence of collisions between its molecules. Such a law of distribution must of course represent a true steady state.

As we shall see later (§ 189), a gas in which abundant collisions are occurring behaves exactly like the fluid of hydrodynamical theory; at every point there is a pressure of the amount given by equation (3) or (19), namely

$$p = \tfrac{1}{3}\rho C^2 = \frac{\rho RT}{m}.$$

If the gas is in a steady state at a uniform temperature T, variations in this pressure hold the gas at rest against the forces exerted by the external field. As we have supposed these forces to be X, Y, Z per unit mass, the hydrostatic equations which express this are

$$\frac{\partial p}{\partial x} = -\rho X, \text{ etc.,}$$

or, inserting the values of p and X,

$$\frac{RT}{m}\frac{\partial \rho}{\partial x} = -\frac{\rho}{m}\frac{\partial \chi}{\partial x}, \text{ etc.}$$

* See, for example, Jeans, *Astronomy and Cosmogony* (2nd edn.), p. 364.

These three equations have the common integral, well known in hydrodynamical theory,

$$RT \log \rho = -\chi + \text{a constant,}$$

or

$$\rho = Be^{-\chi/RT}, \qquad \qquad \ldots\ldots(39)$$

where B is a constant.

If the steady state of the gas is not to be destroyed by collisions, ρ must vary with the potential χ in the way indicated by this equation.

Now ρ expresses the law of distribution in space alone, whereas f expresses it for both space and velocities. If we are to find a true steady state, it must be by extending the ρ given by formula (39) into a formula for f by the addition of velocity terms in u, v and w. Since ρ is a function of χ only, while f must be a function of

$$\tfrac{1}{2}m(u^2 + v^2 + w^2) + \chi,$$

the obvious extension of formula (39) is

$$f(u, v, w, x, y, z) = ABe^{-\frac{\frac{1}{2}m(u^2+v^2+w^2)+\chi}{RT}},$$

where AB is a new constant, being A times the former constant B.

We shall see later that this is the usual law of distribution for a gas in its steady state. It is customary to write h for $1/2RT$, so that the formula becomes

$$f(u, v, w, x, y, z) = AB\, e^{-hm(u^2+v^2+w^2)-2h\chi}. \qquad \ldots\ldots(40)$$

The two factors B and $e^{-2h\chi}$ express the law of distribution of the molecules in space in accordance with formula (39). The remaining factors

$$A\, e^{-hm(u^2+v^2+w^2)} \qquad \qquad \ldots\ldots(41)$$

must therefore express the law of distribution of velocity components for the molecules at any point of space.

24. This law was first discovered by Maxwell, and so is commonly known as Maxwell's law. It occupies a very central position in the kinetic theory, and will be fully discussed in a later chapter.

A more rigorous proof will also be given. The proof just given fails entirely in mathematical rigour, but is of interest because it provides a physical interpretation of the exponential factor which is the outstanding feature of Maxwell's law. We have seen that this is exactly analogous to the exponential in the well-known formula (39) which expresses the falling off of density of a gas in a field of force. This latter formula shews that the chance of a molecule having high potential energy falls off exponentially with the amount of the potential energy: in the same way Maxwell's law tells us that the chance of a molecule having high kinetic energy falls off exponentially with the amount of the kinetic energy. The two exponentials are of exactly similar origin.

We can see still more clearly into the matter by supposing our gas to be the atmosphere, and the field of force to be that of the earth's gravitation. We may think of a molecule of air which reaches the top of the atmosphere as getting there as the result of a succession of lucky collisions with other molecules, each of which gives it enough potential energy to climb one rung higher in the ladder of the atmosphere. The chance of any molecule experiencing a succession of n lucky hits is proportional to a factor of the form $e^{-\alpha n}$, where α is a constant. Thus we can picture in a general way why the chance of a molecule having potential energy χ is proportional to $e^{-\beta \chi}$, where β is a new constant. This gives a physical insight into the origin of the exponential factor in formula (39).

In the same way we may think of a molecule with an exceptionally large energy E, as having acquired it by a succession of lucky collisions. If $\bar{E}$ is the average kinetic energy of a molecule, the number of lucky collisions needed to give kinetic energy E is proportional to $E - \bar{E}$, and the chance of a molecule having experienced this number of lucky collisions is proportional to $e^{-\beta(E-\bar{E})}$, which again is proportional to $e^{-\beta E}$.

The value of an exponential with negative index, such as e^{-x^2}, falls off very rapidly as x becomes large, but it does not become actually zero until x reaches the value $\pm \infty$. Applying this to formula (41), we see that there are molecules with all values of u, v, w right up to $u, v, w = \pm \infty$, but large values of u, v, w are excessively rare. Still even the largest values are not prohibited by the formula.

To notice a simple consequence of this, we know that projectiles escape altogether from the earth's gravitational field if their velocity of projection exceeds about 11 kilometres per second. In the earth's atmosphere, the average molecular velocity is one of hundreds of metres per second. Thus molecular velocities of 11 kilometres per second are excessively rare, but are not non-existent. Of the molecules in the outermost layer of the atmosphere, a certain small proportion must always have velocities in excess of this. These molecules in effect constitute projectiles which will pass off into space never to return. For this reason, the earth is continually losing its atmosphere, although at an excessively slow rate.

The molecules which are most likely to attain speeds of 11 kilometres a second are those of smallest weight, for these have the highest average velocity. Thus hydrogen will be lost more rapidly than nitrogen, and nitrogen more rapidly than oxygen. A simple calculation shews that any hydrogen in the earth's atmosphere would be speedily lost.

The corresponding velocity for the moon is only about 2·4 kilometres per second, so that when the moon had an atmosphere —as it must have had at some time in the past—the rate of escape must have been comparatively rapid. This is why the moon no longer has an appreciable atmosphere. The atmosphere of the other planets and their satellites may be discussed in the same way, with results which are found always to be in agreement with observation.

The mean value of $u^2 + v^2 + w^2$, which we have denoted by C^2, can be obtained from formula (41) by a simple integration (cf. § 91, below), and is found to be $\dfrac{3}{2hm}$ or $3\dfrac{R}{m}T$. In a similar way, the mean value of the velocity $\sqrt{(u^2 + v^2 + w^2)}$, which we shall denote by $\bar{c}$, is found to be $\dfrac{2}{\sqrt{(\pi hm)}}$, which is 0·921 times C, so that $C = 1·086\bar{c}$.

25. If two gases are mixed, and the mixture has attained its steady state, the temperature must be the same for both, so that h, which is equal to $1/2RT$, must be the same for both. Thus the laws of distribution for the two gases will be of the form

$$A\, e^{-hm(u^2+v^2+w^2)}, \qquad \qquad \quad ......(42)$$

$$A'\, e^{-hm'(u'^2+v'^2+w'^2)}, \qquad \qquad \ldots\ldots(43)$$

where m' and u', v', w' denote the mass and components of velocity of a molecule of the second gas.

If we introduce new velocities u'', v'', w'' defined by

$$u' = \sqrt{\frac{m}{m'}}\, u'', \text{ etc.,}$$

this second law may be put in the form

$$A'\, e^{-hm(u''^2+v''^2+w''^2)},$$

which is identical with (41) except for u'', v'', w'' replacing u, v, w. Thus the molecular velocities in the second gas are uniformly $\sqrt{(m/m')}$ times those in the first gas. As a particular case of this,

$$\overline{m'(u'^2+v'^2+w'^2)} = \overline{m(u^2+v^2+w^2)}, \qquad \ldots\ldots(44)$$

so that the average molecular energy is the same in the two gases, the result already obtained in § 11.

The Free Path

26. We may next glance at some problems connected with the free paths of the molecules—i.e. the distances they cover between successive collisions.

Let us fix our attention on any one molecule, which we may call A, and consider the possibility of its colliding with any one of the other molecules $B, C, D, \ldots$. If, for the present, we picture each molecule as a sphere of diameter σ, then a collision will occur whenever the centre of A approaches to within a distance σ of the centre of B, C, D, or any other molecule. Thus we may imagine A extended to double its radius by a sort of atmosphere, and there will be a collision whenever the centre of $B, C, D, \ldots$ passes into this atmosphere (see fig. 7).

Now as A moves in space, we may think of it as pushing its atmosphere in front of it, and for each unit distance that A moves, the atmosphere covers a new volume $\pi\sigma^2$ of space. If there are ν molecules per unit volume, the chance that this volume shall enclose the centre of one of the molecules $B, C, D, \ldots$ is $\pi\nu\sigma^2$. If the molecules $B, C, D, \ldots$ were all standing still in space, the average distance that A would have to travel before it experienced a collision would be

$$\frac{1}{\pi\nu\sigma^2}.$$

This calculation reproduces the essential features of the free-path problem, but is generally in error numerically because it treats the molecule A as being in motion while all the other molecules stand still to await its coming. Thus it gives accurately the free path of a molecule moving with a speed infinitely greater

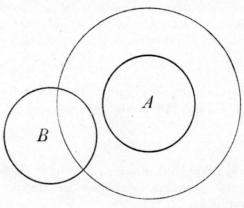

Fig. 7

than that of the other molecules (§ 114). It also gives very approximately the free path of an electron in a gas, because this, owing to its small mass, moves enormously faster than the molecules of the gas. It must however be noticed that σ no longer represents the diameter either of a molecule or of an electron, but the arithmetic mean of the two, and as the diameter of the electron is very small, σ is very nearly the *radius* of the molecule.

In a variety of cases numerical adjustments are needed which will be investigated later (§§ 108 ff.). When all the molecules move with speeds which conform to Maxwell's law, the average free path is found to be

$$\frac{1}{\sqrt{2}\pi\nu\sigma^2}. \qquad\qquad \ldots\ldots(45)$$

Viscosity

27. The free path in a gas which is not in a uniform steady state is of special interest. A molecule which describes a free path PQ of length λ with x-velocity u may be regarded as transporting x-momentum mu from P to Q. If the gas is in a steady state, we may be sure that before long another molecule will transport an equal amount mu of momentum from Q to P, so that, on balance, there is no transfer of momentum.

Suppose, however, that the gas is not in a steady state, but is streaming in parallel layers, with a velocity which varies from one layer to another. To make a definite problem, suppose that the gas is streaming at every point in a direction parallel to the axis Ox, that its velocity of streaming is zero in the plane $z = 0$, and gradually increases as we pass to positive values of z (fig. 8). Let the velocity of streaming at any point be denoted by $\bar{u}$, which is of course the average value of u for all the molecules at this point.

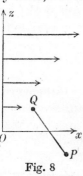

Fig. 8

Again every molecule which describes a free path PQ with velocity u transports momentum mu from P to Q. If the free path crosses the plane $z = 0$, the molecule transports momentum across the plane $z = 0$.

The molecules which cross this plane from the positive to the negative side have, on the average, positive values of u, because they start from regions where the gas is streaming in the direction for which u is positive, and so transport a positive amount of momentum across the plane $z = 0$ into regions where the momentum is negative. But the molecules which cross in the reverse direction start from regions in which u is prevailingly negative, and so on the average carry a negative amount of momentum across the plane into regions where the momentum is positive. Thus both kinds of free path conspire to equalise the momentum, and so also the velocities, on the two sides of the plane. It is as though there were a force acting in the plane $z = 0$ tending to reduce the differences of velocity on the two sides of this plane. This provides the kinetic theory explanation of gas-viscosity.

A rough calculation will give an idea of its amount. Let a typical molecule describe a free path with momentum parallel to Ox of amount mu, and let us suppose the projection of the free path on the axis of z to be λ'. Then the velocities of streaming differ at its two ends by an amount

$$\lambda' \frac{d\bar{u}}{dz},$$

and if the momentum mu is that appropriate to P it is inappropriate to Q by an amount

$$m\lambda' \frac{d\bar{u}}{dz}. \qquad \qquad \ldots \ldots (46)$$

If there are ν molecules per unit volume, and we suppose each to move with an average velocity $\bar{c}$, the number of molecules which cross unit area of the plane $z = 0$ in both directions in unit time is of the order of magnitude of $\nu\bar{c}$; multiplying this by expression (46), we find that the transfer of momentum per unit area per unit time across the plane $z = 0$ would be

$$\nu\bar{c} \times m\lambda' \frac{d\bar{u}}{dz} = \rho\bar{c}\lambda' \frac{d\bar{u}}{dz}$$

if each free path had the same projection λ' on the axis of z. If the free paths are distributed equally in all directions in space and of average length λ, we easily find that the average value of λ' is $\frac{1}{2}\lambda$. Inserting this value for λ' into the formula just obtained, we find that the transfer of momentum is

$$\tfrac{1}{2}\rho\bar{c}\lambda \frac{d\bar{u}}{dz}.$$

The theory of viscosity tells us that the transfer is

$$\eta \frac{d\bar{u}}{dz},$$

where η is the coefficient of viscosity. Thus

$$\eta = \tfrac{1}{2}\rho\bar{c}\lambda. \qquad \qquad \ldots \ldots (47)$$

This simple calculation needs innumerable adjustments, but more exact analysis to be given later shews that the actual value of the coefficient of viscosity does not differ greatly from that just found.

Conduction of Heat

28. A moving molecule transports energy and mass as well as momentum. A molecule which describes the free path PQ of fig. 8 with energy E may be regarded as transporting energy E from P to Q. If the mean energy $\bar{E}$ at P is equal to that at Q, this transfer of energy will be equalised by another in the opposite direction. But if the gas is arranged in layers of equal temperature parallel to the plane of xy, the two mean energies will not be equal, and the first molecule carries, on the average, an excess of energy

$$\lambda' \frac{d\bar{E}}{dz}$$

to Q, while the second carries a deficiency of equal amount to P. Just as in our earlier discussion of momentum, we find that across unit area of any plane there is a transfer of energy equal to

$$v\bar{c}\lambda' \frac{d\bar{E}}{dz}, \quad \text{or} \quad \tfrac{1}{2}v\bar{c}\lambda \frac{d\bar{E}}{dz}. \qquad \ldots\ldots(48)$$

Now the theory of conduction of heat tells us that the transfer of energy is

$$Jk \frac{dT}{dz}, \qquad \ldots\ldots(49)$$

where k is the coefficient of conduction of heat and J is the mechanical equivalent of heat. Comparing expressions (48) and (49), we find as the value of k,

$$k = \frac{v\bar{c}\lambda}{2J} \frac{d\bar{E}}{dT}. \qquad \ldots\ldots(50)$$

We have found as the values of η (the coefficient of viscosity) and C_v (the specific heat at constant volume)

$$\eta = \tfrac{1}{2}\rho\bar{c}\lambda,$$

$$C_v = \frac{1}{Jm} \frac{d\bar{E}}{dT}.$$

Using these relations, equation (50) may be put in the form

$$k = \eta C_v. \qquad \qquad \ldots\ldots(51)$$

The three quantities k, η and C_v which enter into this equation can all be determined experimentally, and the accuracy with which they satisfy the equation provides a test of the truth of our theory. We shall see in Chapter VII that the equation is satisfied with very fair accuracy; the relatively small discrepancies are such as can be ascribed to the simplifying assumptions we have made and to the fact that molecules have more complexity of structure than our theory has allowed for.

29. Finally, if we regard the moving molecules of a gas as transporters of mass, we arrive at a theory of gaseous diffusion, which will be discussed in due course.

Numerical value of the Free Path

30. Of the quantities which enter into the viscosity equation

$$\eta = \tfrac{1}{2}\rho\bar{c}\lambda,$$

we have seen how to calculate $\bar{c}$, and η and ρ can be determined experimentally. Thus λ can be deduced from this equation, or, better, from the more exact equations to be given later. We shall find that the free path in ordinary air is about 6×10^{-6} cm. or a 400,000th part of an inch. In hydrogen, under similar conditions, the free path is about $1 \cdot 125 \times 10^{-5}$ cm. The length of mean free path, as is clear from formula (45), is independent of the temperature or speed of molecular motion, and depends only on the density of the gas. If we double the number of molecules in our gas, keeping its volume unaltered, each molecule has twice as many molecules to collide with, so that at each instant the chance of collision is twice as great as before. Consequently the mean free path is halved. In general the length of the mean free path is inversely proportional to the number of molecules per cubic centimetre of gas. If the pressure is reduced to half a millimetre of mercury, the gas has only $1 : 1520$ of normal atmospheric density, so that the free path in air is about a tenth of a millimetre. In interstellar space, the density may be as low as 10^{-24} gm. per c.c. and the free path is of the order of 10^{16} cm., or 100,000 million km. In internebular space, where the density of gas may be as low as 10^{-29} gm. per c.c., the free path will be of the order of 10^{16} km., or about 1000 light-years.

Comparing the values just obtained for the length of the free path with those previously given for the velocity of motion, we find that the mean time of describing a free path is about $1\cdot3 \times 10^{-10}$ seconds in air under normal conditions, about 2×10^{-7} seconds in air at $0°$ C. at a pressure equal to that of half a millimetre of mercury, and about a thousand million years in internebular space—out here a molecule travels for a long while before meeting another!

As a consequence of these long free paths, molecules out in interstellar space can exist in conditions which are impossible under the far more crowded conditions of laboratory samples, with the result that lines and bands are observed in nebular spectra which cannot be reproduced in the laboratory, although theoretical calculations shew that they ought to occur in gases at extremely low densities.

Molecular Diameters

31. Having estimated the free path from an experimental determination of the coefficient of viscosity, we may proceed to estimate the molecular diameter from the relation

$$\lambda = \frac{1}{\sqrt{2}\pi v\sigma^2}.$$

We find (§ 152) that the diameter of a molecule of air is about $3\cdot75 \times 10^{-8}$ cm., while that of a molecule of hydrogen is only about $2\cdot72 \times 10^{-8}$ cm.

Thus in gas at atmospheric pressure the mean free path of a molecule is some hundreds of times its diameter (160 times for air, 400 times for hydrogen). When the pressure is reduced to half a millimetre of mercury, the free path is hundreds of thousands times the diameter. It is generally legitimate to suppose, as a first approximation, that the linear dimensions of molecules are small in comparison with their free paths.

More definite figures for the lengths of free paths and the size of molecules will be given later, but it is no easy matter to discuss these questions with any precision; we cannot even define the size of a molecule. The difficulty arises primarily from our ignorance of the shape and other properties of the molecule. If molecules were in actual fact elastic spheres, the question would be simple enough; the size of the molecule would be measured simply by the diameter of the sphere. The molecules are not, however, elastic spheres; if they are *assumed* to be elastic spheres, experiment leads to discordant results for the diameters of these spheres, shewing that the original assumption is unjustifiable. Not only are the molecules not spherical in shape, but also they are surrounded by fields of force, and most experiments measure the extension of this field of force, rather than that of the molecules themselves.

These questions will be discussed more fully at a later stage (§§ 146 ff.).

Chapter III

PRESSURE IN A GAS

CALCULATION OF PRESSURE IN AN IDEAL GAS

32. Leaving our general survey, we now turn to a more detailed study of various special problems. The present chapter will be concerned with the calculation of the pressure in a gas.

This has already been calculated on the supposition that the molecules of the gas are hard spheres of infinitesimal size, which exert no forces on one another except at collisions. We now give an alternative calculation, which still assumes the molecules to be infinitesimal in size and exerting no forces except at collisions, although not necessarily hard or spherical.

33. Let dS in fig. 9 represent a small area of the boundary of a vessel enclosing the gas. Let there be ν molecules per unit volume of the gas, and let these be divided into classes, so that all the molecules in any one class have approximately the same velocities, both as regards magnitude and direction. Let $\nu_1, \nu_2, \ldots$ be the numbers of molecules in these classes, so that

$$\nu_1 + \nu_2 + \ldots = \nu.$$

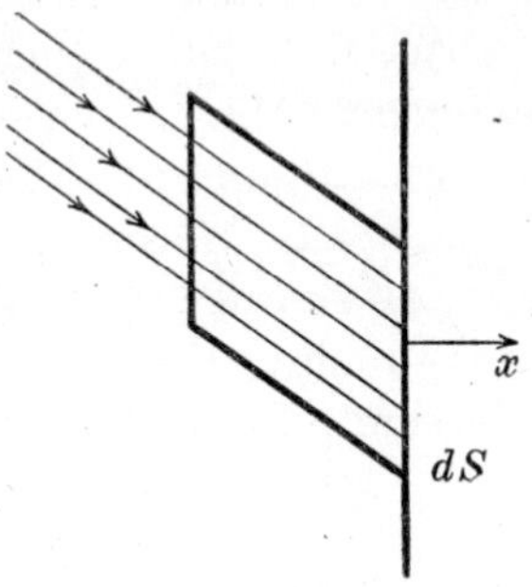

Since the molecules of any one class all move in parallel directions, they may be regarded as forming a shower of molecules of density ν_1 per unit volume, in which every molecule moves with the same velocity. Some showers will be advancing towards the area dS of the boundary, some receding from it. Let us suppose that the shower formed by molecules of the first class is advancing.

The molecules of this shower which strike dS within an interval of time dt will be those which, at the beginning of the interval, lie within a certain small cylinder inside the vessel (see fig. 9).

Fig. 9

This cylinder has dS for its base, and its height is $u_1 dt$, where u_1 is the velocity of molecules of the first shower in the direction normal to dS, which we shall take to be the axis of x. Thus the volume of the cylinder is $u_1 dt dS$, and the number of molecules of the first shower inside it is $\nu_1 u_1 dt dS$.

Each of these molecules brings to dS momentum mu_1 in a direction normal to the boundary. Thus in time dt the whole shower brings momentum $m\nu_1 u_1{}^2 dt dS$, and all the showers which are advancing on dS bring an aggregate momentum $\Sigma m\nu_1 u_1{}^2 dt dS$, where the summation is over all showers for which u is positive.

The pressure between the boundary and the gas first reduces this x-momentum to zero, and then creates new momentum in the opposite direction (u negative).

The pressure which, acting steadily for a time dt, will reduce the original momentum to zero is $\Sigma m\nu_1 u_1{}^2 dS$, where the summation is again over all showers for which u is positive.

The pressure needed to create the new momentum after impact can be calculated in precisely the same way. We enumerate the molecules which have impinged on dS within an interval dt, and find that their momentum would be produced by a steady pressure $\Sigma m\nu_1 u_1{}^2 dS$, where the summation is now over all showers for which u is negative.

Combining these two contributions, we find that the total pressure $p\,dS$ on the area dS is given by

$$p\,dS = \Sigma m\nu_1 u_1^2 dS, \qquad \text{......(52)}$$

where the summation is over all the showers of molecules.

The value of $\Sigma \nu_1 u_1^2$ is the sum of the values of u^2 for all the molecules in unit volume, and this is equal to $\nu \overline{u^2}$. Thus equation (52) may be written in the form

$$p = m\nu \overline{u^2} = \rho \overline{u^2}. \qquad \text{......(53)}$$

As we have seen (§ 12) that

$$m\overline{u^2} = m\overline{v^2} = m\overline{w^2} = \tfrac{1}{3}mC^2 = RT, \qquad \text{......(54)}$$

equation (53) again assumes the new forms

$$p = \tfrac{1}{3}\rho C^2 = \nu RT. \qquad \text{......(55)}$$

If the gas consists of a mixture of molecules of different kinds, the summation of equation (52) must be extended to all the types of molecules, and the final result, instead of equation (55), is

$$p = (\nu + \nu' + \ldots) RT. \qquad \ldots\ldots(56)$$

If a volume v of homogeneous gas contains N molecules in all, then $(\nu + \nu' + \ldots) v = N$, and equations (55) and (56) may be combined in the single equivalent equation

$$pv = NRT. \qquad \ldots\ldots(57)$$

34. In making this calculation, it has not been necessary to make any assumption as to the way in which momenta or velocities are distributed in the various showers after reflection from the boundary.

Certain experiments by Knudsen* have led him to the conclusion that the directions of molecular motion after impact are not in any way related to the directions of motion before impact, but that the molecules start out afresh after impact in random directions from the boundary, so that the number which make an angle between θ and $\theta + d\theta$ with the normal to the surface will be proportional to $\sin\theta\, d\theta$. If this were

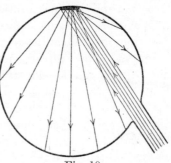

Fig. 10

the true law, it is easily seen that when molecules are reflected from a minute area of the inner surface of a sphere, the number which would strike any area of the sphere would be proportional simply to the area. Knudsen has tested this by reflecting molecules from a small area of a glass sphere which was kept at room temperature, and allowing them, after reflection, to form a deposit on the remainder of the sphere which was kept at liquid air temperature (fig. 10). The law was found to be well obeyed.

From the same supposition, Knudsen has deduced certain other properties which he finds† to be in good agreement with

* *Ann. d. Phys.* **48** (1915), p. 111, and *The Kinetic Theory of Gases* (Methuen, 1934), pp. 26 ff. † *Ann. d. Phys.* **48** (1915), p. 1113.

observations on the flow of gases. The assumption in question has also been tested by R. W. Wood,* who finds that it agrees very closely with experiment. The exact law of reflection, although immaterial for the calculation of normal pressure, is of importance when tangential stresses have to be estimated, as in the flow of gases through tubes (§§ 143, 144 below).

Langmuir† and others have supposed that when a gas is in contact with a solid, those molecules of the gas which collide with the surface of the solid are either partially or wholly adsorbed by the solid, and subsequently re-emitted by the surface in a manner which does not depend on their history before adsorption took place; the molecule, so to speak, ends one life when it strikes the boundary and after an interval starts out again on a new existence. This supposition has obtained considerable experimental confirmation and provides a rational explanation of Knudsen's law. A great deal of work has been done on the subject, but many complications and difficulties remain.‡

35. As we have already noticed (§§ 13, 14), equation (57) predicts all the well-known laws of gases—the laws of Avogadro, Dalton, Boyle and Charles. These laws have of course only been shewn to be valid within the limits imposed by the assumptions made in proving them. The principal of these assumptions have been that the molecules (or other units by which the pressure is exerted) are so small that they may be treated as points in comparison with the scale of length provided by intermolecular distances, and that the forces between molecules are negligible except at collisions. Thus the laws are best regarded as ideal laws, which can never be absolutely satisfied, but which are satisfied very approximately in a gas of great rarity. An imaginary gas in which the molecules exert no forces except at collisions, and have dimensions so small in comparison with the other distances involved that they may be regarded as points is spoken of as an "ideal" gas. The foregoing laws are always true

---

* *Phil. Mag.* **30** (1915), p. 300; see also *Phil. Mag.* **32** (1916), p. 364.

† *Phys. Rev.* **8** (1916), p. 149; *Journ. Amer. Chem. Soc.* **40** (1918), p. 1361; *Trans. Faraday Soc.* **17** (1921), Part III.

‡ For a good summary, see Loeb, *Kinetic Theory of Gases* (2nd edn., 1934), pp. 338 ff.

for an ideal gas; for real gases they will be true to within varying degrees of closeness, the accuracy of the approximation depending on the extent to which the gas approaches the state of an ideal gas.

The method of evaluating the pressure which has been given in § 33 in no way requires that the medium should be gaseous, although the resulting laws of Dalton, Boyle and Charles are usually thought of only in relation to gases. Clearly, however, these laws must apply to any substance with a degree of approximation which will depend only on the nearness to the truth of the assumptions just referred to.

In point of fact the laws are found to be true (as they ought to be) for the osmotic pressure of weak solutions. They are also true for the pressure exerted by free electrons moving about in the interstices of a conducting solid, and also for the pressure exerted by the "atmosphere" of electrons surrounding a hot solid. Each of these pressures p may be assumed to be given by formula (55), where ν is the number of free electrons per unit volume.

Numerical Estimate of Velocities

36. From equation (55) we at once obtain the relation

$$C^2 = \frac{3p}{\rho} \qquad \qquad \ldots\ldots(58)$$

which, as we have already noticed, provides the means of estimating the molecular velocity of a gas, as soon as corresponding values of p and ρ have been determined observationally.

The density of oxygen at 0° and at the standard atmospheric pressure of $1\cdot01323 \times 10^6$ dynes per sq. cm. is found to be $0\cdot0014290$, whence we can calculate from equation (58) that

$$C = 461\cdot2 \text{ metres per second.}$$

Also we have seen (§ 24) that $C = 1\cdot086$ times $\bar{c}$, the mean velocity of all molecules, so that

$$\bar{c} = 425 \text{ metres per second.}$$

We have also obtained the relation

$$C^2 = \frac{3RT}{m}, \qquad \qquad \ldots\ldots(59)$$

where T is the temperature on the absolute scale, and R is a constant which is the same for all gases. Thus as between one kind of molecule and another, C varies as $m^{-\frac{1}{2}}$. Knowing the values of C and $\bar{c}$ for oxygen at $0°$, it is easy to calculate them for any other substance. It is also easy to calculate C and $\bar{c}$ for other temperatures since equation (59) shews that both vary as $T^{\frac{1}{2}}$, the square root of the absolute temperature.

37. In this way the following table has been calculated:

Molecular Velocities at 0° C.

Gas (or other substance)	Molecular weight (O = 16)	$\dfrac{R}{m}$	C (cm. per sec. at 0° C.)	$\bar{c}$ (cm. per sec. at 0° C.)
Hydrogen	2·016	4127×10^4	1839×10^2	1694×10^2
Helium	4·002	2077×10^4	1310×10^2	1207×10^2
Water-vapour	18·016	462×10^4	615×10^2	565×10^2
Neon	20·18	412×10^4	584×10^2	538×10^2
Carbon-monoxide	28·00	297×10^4	493×10^2	454×10^2
Nitrogen	28·02	297×10^4	493×10^2	454×10^2
Ethylene	28·03	297×10^4	493×10^2	454×10^2
Nitric oxide	30·01	277×10^4	476×10^2	438×10^2
Oxygen	32·00	260×10^4	461×10^2	425×10^2
Argon	39·94	208×10^4	431×10^2	380×10^2
Carbon-dioxide	44·00	189×10^4	393×10^2	362×10^2
Nitrous oxide	44·02	189×10^4	393×10^2	362×10^2
Krypton	82·17	101×10^4	286×10^2	263×10^2
Xenon	131·3	63×10^4	227×10^2	209×10^2
Mercury vapour	200·6	$41·6 \times 10^4$	185×10^2	170×10^2
Air	—	$[287 \times 10^4]$	485×10^2	447×10^2
Free electron	$\frac{1}{1835}$ (H = 1)	$1·515 \times 10^{11}$	$1·114 \times 10^7$	$1·026 \times 10^7$

We find that for oxygen $R/m = 259·6 \times 10^4$, while the value of m is found, as in § 19, to be $53·12 \times 10^{-24}$ grammes. Hence, by multiplication,

$$R = 1·379 \times 10^{-16}. \qquad \ldots\ldots(60)$$

This quantity is a universal constant, depending only on the particular scale of temperature employed. It will be remembered that $\frac{3}{2}R$ is the kinetic energy of translation of any molecule whatever at a temperature of $1°$ absolute (cf. equation (17)). It is often convenient to denote this by a single symbol α, so that

$$\alpha = \tfrac{3}{2}R = 2·068 \times 10^{-16}. \qquad \ldots\ldots(61)$$

The kinetic energy of translation of a molecule (or free atom or electron) at a temperature of T degrees absolute is now αT.

Other numerical values which are frequently of service are

$$RT_0 = 3 \cdot 767 \times 10^{-14}, \quad \alpha T_0 = 5 \cdot 651 \times 10^{-14},$$

where $T_0 = 273 \cdot 2°$ (centigrade), the temperature of melting ice (0° C.) on the absolute scale.

38. The order of magnitude of the molecular velocities just obtained might have been foreseen, without detailed calculation, in the following way.

Any disturbance in a gas will of course first produce an effect on the molecules in its immediate neighbourhood. When these molecules collide with those in the adjacent layers of gas, the effect of the disturbance is carried on into that layer, and so on indefinitely. Thus the molecules of a gas act as carriers of any disturbance, and the disturbance is propagated through the gas with a velocity comparable with the mean velocity of motion of the molecules, just as, for instance, news which is carried by relays of messengers spreads with a velocity comparable with the mean rate at which the messengers travel. The propagation of a disturbance in the gas is, however, nothing but the passage of a wave of sound, whence we see that the mean molecular velocity in any gas must be comparable with the velocity of sound in the same gas. Actually the velocity of sound a is given by the well-known hydrodynamical formula

$$a^2 = \gamma \frac{p}{\rho},$$

where γ is the ratio of the two specific heats of the gas in question (cf. § 222 below). On replacing p by its value, $\frac{1}{3}\rho C^2$, this equation becomes

$$a = \sqrt{\tfrac{1}{3}\gamma} C.$$

For diatomic gases at ordinary temperatures, $\gamma = 1\frac{2}{5}$, so that

$$a = 0 \cdot 683 C = 0 \cdot 742 \bar{c}, \qquad \qquad \ldots\ldots(62)$$

shewing the actual relation between the velocity of sound and the velocities C and $\bar{c}$ for these gases.

For instance, the table gives for air at 0° C., $C = 485$ metres per second, whence formula (62) leads to $a = 331 \cdot 3$ metres per second for the velocity of sound in air, which approximates very closely to the true value.

Effusion of Gases

39. The general order of magnitude of molecular velocities can also be discovered experimentally by making a minute hole in the wall of a containing vessel and allowing the imprisoned gas to stream out; the velocity of efflux is nothing else than the velocities of the individual molecules. These would have been simply molecular velocities inside the vessel had the hole not been present. The hole in the vessel causes the molecular velocities to persist out into space, where they can be measured observationally. In fig. 9 (p. 51) imagine that the element of surface dS is replaced by a minute trap-door, which is suddenly opened. The number of molecules which stream through the trap-door in time dt is of course equal to the number which would have impinged on the element dS of the boundary had the trap-door remained closed. If each molecule were moving directly towards the trap-door with a velocity $\bar{c}$, the mean velocity of all the molecules of the gas, this number would be

$$\nu \bar{c}\, dS\, dt,$$

so that the rate of efflux, measured in terms of mass per unit time per unit area of aperture, would be

$$m\nu\bar{c} = \rho\bar{c}. \qquad \qquad \dots\dots(63)$$

Actually half of the molecules are moving away from the aperture, so that this formula must be reduced by a factor of $\frac{1}{2}$, and those which are moving towards the aperture are moving in all directions at random, and therefore according to the law $\sin\theta\, d\theta\, d\phi$. A simple integration shews that this reduces formula (63) by a further factor of $\frac{1}{2}$. When both of these considerations are taken into account, formula (63) must be replaced by

$$\tfrac{1}{4}\rho\bar{c}.$$

Thus the rate of efflux is the same as if the whole gas of density ρ streamed out of the aperture with a uniform velocity $\frac{1}{4}\bar{c}$.

Using the value for $\bar{c}$ obtained in § 24, we find that the formula can also be put in the form

$$\rho\sqrt{\frac{RT}{2\pi m}}, \qquad \ldots\ldots(64)$$

shewing that *the rates of efflux of different gases at the same density and temperature vary inversely as the square roots of the molecular weights of the gases.*

For gases at the same pressure and temperature, the densities are proportional to the molecular weights, so that the rate of efflux varies as the square root of either. This of course follows more directly from the equipartition of energy (§ 11).

40. In 1846 Graham[*] made some experiments to test this latter law, measuring the speeds of efflux of various gases coming through fine perforations in a brass plate. The results are shewn in the following table:

Efflux of Gases

Gas	√(density) (air = 1)	Rate of efflux (air = 1)
Hydrogen	0·263	0·276
Marsh gas	0·745	0·753
Ethylene	0·985	0·987
Nitrogen	0·986	0·986
Air	1·000	1·000
Oxygen	1·051	1·053
Carbon-dioxide	1·237	1·203

Had the law been completely confirmed, the numbers in the second and third columns would have been identical; we should then have had a direct experimental proof of the law of equipartition of energy. Actually the numbers given in the table seem rather to suggest that the law is not obeyed with any great accuracy.

Knudsen found the reason for this in 1909;[†] it is simply that as the molecules are not mere points of infinitesimal size, they collide with one another and affect one another's motion

[*] *Phil. Trans. Roy. Soc.* **136** (1846), p. 573.
[†] *Ann. d. Phys.* **28** (1909), p. 75.

in passing through the perforations. He performed experiments similar to those of Graham, but used a hole only 0·025 mm. diameter, cut in a platinum strip of only 0·0025 mm. thickness. He worked down to pressures as low as 0·01 mm. of mercury, at which gaseous collisions are very rare indeed, and found that the theoretical laws were well obeyed so long as the mean free path in the issuing gas was at least ten times the diameter of the hole. As soon as the mean free path became less than this, rather more gas came out than was predicted by the formula, and as the pressure was still further increased, the efflux of gas gradually passed over to that predicted by the ordinary hydrodynamical formula for the flow of fluid through a hole.

A number of experimenters have tried to determine the molecular weights of various gases by a use of formula (64), but Knudsen's results make it clear that accurate results cannot be expected unless extreme precautions are taken.

Formula (64) is applicable only to a gas flowing out into a perfect vacuum. If there is any gas on the farther side of the orifice, some of the molecules of the issuing gas will collide with the molecules of the external gas and be driven back, thus reducing the rate of efflux. If, however, the external gas is of very low density, there will be few collisions of this kind, and formula (64) will still give a good approximation to the rate of efflux.

Transpiration of Gases

41. For experimental purposes, it is often better, instead of using a single orifice or perforation, to use the large number of very small orifices provided by the interstices in a plug of porous material, such as unglazed earthenware or meerschaum. The phenomenon is then spoken of as "transpiration" rather than "effusion".

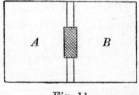

Fig. 11

Imagine a vessel of gas divided into two parts by such a porous plug (fig. 11). Transpiration or effusion will be going on from each side of this plug to the other. If the two chambers into which the vessel is divided are denoted by A and B, there will be some gas from A crossing through the porous plug into B, and similarly some from B crossing into A. If the pressures in both chambers are sufficiently low, both rates of transpiration may, as an approximation, be supposed given by formula (64). If the gases in the two chambers are the same in all respects, the two rates of effusion will of course be the same. If, however, one chamber is kept warmer than the other, then the rates of effusion will not be the same, and we have the phenomenon of thermal transpiration.

Let T_A, T_B be the temperatures of the two chambers, and let the corresponding densities and pressures be ρ_A, ρ_B and p_A, p_B. If the temperature difference is permanently maintained, the flow of gas will continue until a state is attained in which the flow from A to B is exactly equal to that from B to A. Formula (64) shews that this state will be reached when

$$\rho_A \sqrt{T_A} = \rho_B \sqrt{T_B}. \qquad \dots\dots(65)$$

The ratio of the pressures p_A, p_B is accordingly given by

$$\frac{p_A}{p_B} = \frac{\rho_A T_A}{\rho_B T_B} = \sqrt{\frac{T_A}{T_B}}. \qquad \dots\dots(66)$$

Thus if the two chambers are kept unequally heated, a flow of gas will be set up, and will continue until the difference of pressure between the two sides is established which is expressed by equation (66).

Osborne Reynolds* investigated this phenomenon in a series of experiments in which the two chambers were kept at temperatures of 8° C. and 100° C. When a steady state was attained, the pressures were measured, and it was found that, so long as the pressure was sufficiently low, equation (65) was satisfied with very considerable accuracy. At higher pressures this equation failed, as was to be expected.

If the chambers A and B in fig. 11 are not only connected by the porous plug, but also by an external pipe, which keeps the pressures in A and B equal, then a steady state cannot be attained so long as the temperatures are kept permanently at different temperatures T_A, T_B. Instead of this, there will be a steady flow of gas through the cycle formed by the chambers

* *Phil. Trans.* **170**, II (1879), p. 727.

A, B and the pipe, a flow which is suggestive of and analogous to that of a thermoelectric current.

Cohesion of Gases

42. An entirely different situation occurs if the chamber B in fig. 11 contains no gas, while chamber A is filled with gas kept at temperature T_A. This gas will flow through the plug or orifice into the chamber B, and its temperature as it arrives in the chamber B, say T_B, could be measured by a thermometer placed in B.

Suppose that the molecules of the gas had been held together by strong forces of cohesion, so that each molecule was attracted by the other molecules of the gas, or at least by those in its immediate proximity. Then each molecule, while passing through the plug, would be under an attraction towards the molecules in the chamber A, and as this attraction would reduce its velocity, the average velocity of molecules arriving in B would be less than the average velocity of molecules in A; the temperature T_B would be less than T_A.

Thus we see that an examination of the temperature of a gas after transpiration or effusion will give us information as to the existence or non-existence of forces of cohesion in a gas. Experiments of this general type had been devised and conducted by Gay-Lussac in 1807 and Joule in 1845, although they had used merely a tube and a tap in place of the porous plug used by the later experimenters. The more sensitive form of apparatus was devised by Lord Kelvin, who carried out a delicate and crucial set of experiments in collaboration with Joule.*

The earlier experiments had failed to detect any temperature change in the gas, shewing that the forces of cohesion in a gas were at least small. In the more elaborate experiments of Joule and Kelvin, slight falls of temperature were observed; for instance, in an experiment in which air passed by transpiration from a pressure of about four atmospheres to a pressure of one atmosphere, the change of temperature observed was a fall of $0 \cdot 9° C$. In general it was found that for air and many of the more

---

* The original papers will be found in the *Phil. Trans. Roy. Soc.* (**143**, p. 357; **144**, p. 321; **150**, p. 325 and **152**, p. 579). See also Lord Kelvin's *Collected Works*, **1**, p. 333.

permanent gases the cooling, although appreciable, was very slight, while for hydrogen a slight heating was observed. This change of temperature was known as the Joule-Thomson effect. It did not, as was at first thought, establish the existence of a force of cohesion in gases—or of negative cohesion in hydrogen. For as the gas passed through the porous plug—generally a wad of cotton-wool—the pressure fell, so that the gas actually going through the plug was doing work in pushing forward, by its pressure, the less dense gas in front of it, while work of the same kind but greater in amount was done on it by the denser gas behind. The net result was of course a gain of energy to the gas traversing the plug, so that even if this gas consisted of infinitesimal billiard balls which exerted no cohesive forces on one another, it would still emerge with an increase of energy, and this would necessarily appear as an increase of temperature when equilibrium had been attained (cf. § 92 below).

When the observed results are corrected to allow for this, it is found that the molecules of all gases, including hydrogen, lose kinetic energy in passing through the plug. The loss for air is found to be 0·051 calorie per gramme for each atmosphere drop of pressure; the corresponding figure for hydrogen is 0·06 calorie per gramme. It follows that the molecules of a gas may not properly be treated as points exerting no forces on one another; they attract one another, so that there are forces of cohesion in a gas, just as there are in a liquid or a solid.

CALCULATION OF PRESSURE IN A NON-IDEAL GAS

43. In §33 the pressure in a gas was calculated on the assumptions that the molecules were infinitesimal in size, and that they exerted no forces on one another except when they were actually in collision. We now see that neither of these assumptions is true for an actual gas, and so must proceed to calculate the pressure for a real gas in which the molecules are of finite size, and exert forces of cohesion on one another even when they are not in contact. Many such calculations have been made, the most famous being that of Van der Waals (1873) to which we now turn.

Van der Waals' Equation

44. We shall face our difficulties separately, and so begin by supposing that the molecules are of finite size, being spheres of finite diameter σ, but that these exert no forces on one another except when in actual contact.

Let there be N molecules A, B, C, ... in a volume v, and imagine, as before, that a sphere of radius σ is drawn round the

centre of each of these molecules (fig. 12). Since the centres of two molecules cannot be within a distance less than σ from each other, it is impossible for the centre of molecule A to lie within any of the spheres of radius σ surrounding the $(N-1)$ other molecules B, C, D, ... Each of these spheres has a volume $\frac{4}{3}\pi\sigma^3$, so that their combined volume is $\frac{4}{3}(N-1)\pi\sigma^3$. Since N is a very great number, we may neglect the difference between $N-1$ and N and write this total volume in the form $\frac{4}{3}N\pi\sigma^3$.

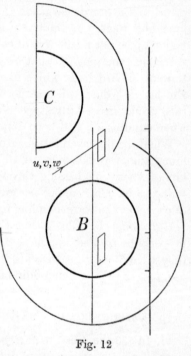

Fig. 12

We shall proceed on the supposition that this total volume is only a small fraction of the total volume v of the gas. In this case we may disregard the possibility of two of the spheres of radius σ intersecting, and so suppose that the space outside the spheres is of volume $v - \frac{4}{3}N\pi\sigma^3$. This is, then, the space available for the centre of molecule A. Thus the chance of finding the centre of molecule A inside a specified volume dv is nil if this volume lies inside one of the spheres, and otherwise is

$$\frac{dv}{v - \frac{4}{3}N\pi\sigma^3}. \qquad \dots\dots(67)$$

Now let molecule A be moving with a velocity of components u, v, w, the component u being, as before, perpendicular to a certain region dS of the boundary. The molecule A will hit this region dS of the boundary within a small interval of time dt if, at the beginning of this interval, its centre lies within a small element of volume $u\,dS\,dt$, which is at a distance $\frac{1}{2}\sigma$ from the boundary. Let us now identify this element of volume with the dv of the last paragraph.

We must now consider the two possibilities separately of this element of volume lying within one of the $N-1$ spheres of radius σ, and of its not so lying.

We can imagine each sphere divided into two hemispheres by a plane through its centre parallel to dS; we may describe these as the nearer and farther hemispheres. Each point in the farther hemisphere is farther from dS than the centre of the sphere, so that no such point can ever be at a distance less than $\frac{1}{2}\sigma$ from the boundary. Since the element of volume is only at a distance $\frac{1}{2}\sigma$ from the boundary, it can never lie in one of the farther hemispheres, but all points in the nearer hemispheres are open to it. The total volume of the farther hemispheres is $\frac{2}{3}(N-1)\pi\sigma^3$, or, again disregarding the difference between $N-1$ and N, is $\frac{2}{3}N\pi\sigma^3$. Thus the chance of dv lying inside one of the spheres is

$$\frac{\frac{2}{3}N\pi\sigma^3}{v},$$

while the chance of its not so lying is

$$\frac{v-\frac{2}{3}N\pi\sigma^3}{v}. \qquad \ldots\ldots(68)$$

In the former case there is no chance of a collision. In the latter case the chance of a collision within the interval dt is, from formula (67),

$$\frac{N\,dv}{v-\frac{4}{3}N\pi\sigma^3}. \qquad \ldots\ldots(69)$$

The chance of a collision is found by multiplying together the two probabilities (68) and (69) and so is

$$\frac{N\,dv}{v}\frac{v-b}{v-2b},$$

where $$b = \tfrac{2}{3}N\pi\sigma^3, \qquad \qquad \ldots\ldots(70)$$

and so is four times the total volume of all the molecules of the gas. Since b has already been supposed to be small in comparison with v, we may neglect $(b/v)^2$, and write this in the form

$$\frac{Ndv}{v-b}.$$

We accordingly see that when b/v is small—i.e. when the aggregate volume of the molecules is small compared with the space occupied by the gas—the effect of the finite size of the molecules is adequately allowed for by replacing v by $v-b$, i.e. by reducing the volume of the vessel by four times the total volume of N molecules. When this is done we have as the equation giving the pressure, in place of equation (57),

$$p(v-b) = NRT. \qquad \qquad \ldots\ldots(71)$$

Since we are neglecting squares of b/v, this may be put in the alternative form

$$pv = NRT\left(1+\frac{b}{v}\right), \qquad \qquad \ldots\ldots(72)$$

which is frequently found useful.

When b/v is not treated as a small quantity, the calculation of the pressure presents a much more serious problem. Boltzmann[*] has carried the calculation as far as squares of $(b/v)^2$, and finds that the pressure is given by

$$pv = NRT\left\{1+\frac{b}{v}+\frac{5}{8}\left(\frac{b}{v}\right)^2 + \ldots\right\},$$

but this is hardly found to agree better with observation than the simpler equation (72).

45. A second correction is necessitated by the forces of cohesion, the existence of which we noted in § 42. So long as a molecule is sufficiently remote from the surface, forces act on it which will vary continually, both in direction and magnitude, but which, when averaged over a sufficient interval of time, are likely to cancel out, and leave an average resultant equal to zero. Thus this second correction will not affect the pressure of the gas

---

[*] *Vorlesungen über Gastheorie,* **2,** § 51.

at internal points; this is given by equation (71). But to calculate the pressure of the gas at or near the boundary, these forces must be taken into account.

Let us fix our attention on a molecule which is as near to the boundary as it can go. If the force from each adjacent molecule is resolved into components in and at right angles to the boundary, all directions in the boundary plane are equally likely for the first component, but the second component is invariably directed inwards. Thus when we average over a sufficient length of time, the resultant force will be a force directed inwards at right angles to the boundary.

Thus the effect of all the intermolecular forces is that of a permanent field of force acting upon each molecule at and near the surface. It is this field of force which is often pictured as giving rise to the phenomena of capillarity and surface-tension in liquids. It can be regarded as exerting a steady inward pull upon the outermost layer of molecules of the gas. In magnitude this pull will be proportional jointly to the number of molecules per unit area in this layer, and to the intensity of the normal component of force. As each of these two factors is directly proportional to the density of the gas, the pull will be proportional to the square of the density; let us suppose that it is of amount $c\rho^2$ per unit area, where c is a constant depending only on the nature of the gas. Then the molecules, when they reach the boundary, are no longer deflected by impact alone, but by the joint action of their impact with the boundary and of this pull. Thus their change of momentum may be supposed to be produced by a total pressure $p + c\rho^2$ per unit area, instead of by the simple pressure p.

Hence equation (71) must be further amended by writing it in the form

$$(p + c\rho^2)(v - b) = NRT,$$

or again, replacing ρ by Nm/v, and putting $cN^2m^2 = a$,

$$\left(p + \frac{a}{v^2}\right)(v - b) = NRT. \qquad \ldots\ldots(73)$$

This is Van der Waals' equation connecting p, v and T. So long as we confine our attention to a single mass of gas, a and b are constants, but in general they depend on the amount of gas as well as on its nature, a being proportional to the square and b to the first power of the amount of gas.

Dieterici's Equation

46. Van der Waals' treatment of the forces of cohesion overlooks the fact that when cohesive forces exist, some molecules which would have reached the boundary had there been no such forces may never reach the boundary at all, being deflected by the cohesion forces before their paths meet the boundary. Such molecules exert no pressure on the boundary, whereas Van der Waals' argument has supposed them to exert a negative pressure. Because of this, equation (73) admits of negative values for p, although an examination of the physical conditions shews that the true value of p must necessarily be positive.

This objection is of no weight so long as it is clearly recognised that equation (73) is true only to the first order, as regards deviations from Boyle's law, but the equation is often, and very usefully, applied to cases where these deviations cannot properly be regarded as small. Because of this, Dieterici* has proposed an alternative equation of state which is not open to this last objection. We have seen that the molecules which are close to the boundary of the gas act as though they were drawn back into the gas by a permanent field of force. Let χ be the work done in drawing a molecule from the interior of the gas to any assigned position near the boundary against these forces, then the density ρ' at this point is, as in equation (39),

$$\rho' = \rho\, e^{-2h\chi},$$

where ρ is the density in the interior, and so is what is ordinarily meant by the density of the gas. We now take this point close up to the boundary of the gas, and find that the pressure on the boundary p is given by

$$p = \rho'\overline{u^2} = \rho\overline{u^2} e^{-2h\chi},$$

where ρ' and χ are the density and potential at the boundary of the gas.

* *Wied. Ann.* **65** (1898), p. 826 and **69** (1899), p. 685.

Thus the field of force reduces the pressure to a value which is less than that given by Boyle's law by a factor $e^{-\chi/RT}$. If we assume that the pressure is similarly reduced when the molecules are of finite size, we obtain the pressure at the boundary in the form

$$p = \frac{NRT}{v-b} e^{-\chi/RT}.$$

The work χ is clearly proportional to ρ, to a first approximation, and is easily seen to be equal to a/Nv, where a is the a of Van der Waals, leading to the equation

$$p = \frac{NRT}{v-b} e^{-a/NRTv}. \qquad \ldots\ldots(74)$$

This equation was first given by Dieterici in 1898, and is found, as we shall see later, to fit the observed facts considerably better than the equation of state of Van der Waals.

The General Calculation of Pressure

47. Logically, neither of the two foregoing pressure calculations is completely satisfactory, as can be seen in the following way. The correction a arises from forces which the molecules exert on one another when reasonably near to one another, i.e. when their centres are at distances somewhat greater than σ apart. The correction b arises from forces which the molecules exert on one another when their centres are at exactly the distance σ apart. It cannot, however, be supposed that the forces acting on natural molecules can be divided into two distinct types in this way. The forces between two molecules must change continuously with the distance, so that the a and b of Van der Waals' equation ought to be different contributions from a more general correction, and so ought to be additive. This is not the case with the equations either of Van der Waals nor of Dieterici.

There is, however, a wider calculation of pressure, first given by Clausius* in 1870, to which this objection does not apply. It proceeds by studying the motions of the molecules of a gas as these move under the influence of perfectly general forces, these

* *Phil. Mag.* Aug. 1870.

including the forces at collision, the forces of cohesion between molecules, and the forces of impact at the boundary. In this way the pressure is found in terms of the general forces between molecules. To this we now turn.

The Virial of Clausius

48. We suppose the motion of any molecule in a gas to be governed by the Newtonian equations

$$X = m\frac{d^2x}{dt^2}, \text{ etc.,} \qquad \ldots\ldots(75)$$

where X, Y, Z are the components of the total force acting on the molecule. Multiplying these three equations by x, y, z respectively and adding, we obtain

$$(xX + yY + zZ) = m\left(x\frac{d^2x}{dt^2} + y\frac{d^2y}{dt^2} + z\frac{d^2z}{dt^2}\right)$$

$$= m\frac{d}{dt}\left(x\frac{dx}{dt} + y\frac{dy}{dt} + z\frac{dz}{dt}\right) - m\left[\left(\frac{dx}{dt}\right)^2 + \left(\frac{dy}{dt}\right)^2 + \left(\frac{dz}{dt}\right)^2\right]$$

$$= \tfrac{1}{2}m\frac{d}{dt}\left[\frac{d}{dt}(x^2+y^2+z^2)\right] - mc^2, \qquad \ldots\ldots(76)$$

where c, as before, is the velocity of the molecule.

As the motion of the gas proceeds, $\frac{d}{dt}(x^2+y^2+z^2)$ continually fluctuates in value, but there is no tendency to a steady increase or decrease. If, then, we sum over all the molecules of the gas, the first term on the right of equation (76) disappears, and we have

$$\tfrac{1}{2}\Sigma mc^2 = -\tfrac{1}{2}\Sigma(xX + yY + zZ). \qquad \ldots\ldots(77)$$

The expression on the right is known as the Virial of Clausius; we see that it is equal to the total kinetic energy of translation of the molecules.

Contributions to the virial are made by all the forces which ever act upon the molecules. We may divide these into

(1) The forces at collisions between molecules.

(2) The forces of cohesion between molecules.

(3) The forces of impact between molecules and the boundary of the containing vessel.

Let us first examine the contribution made by (3); it is through this that the pressure is introduced into our analysis.

Let dS be any element of surface of the containing vessel, the coordinates of its centre being x, y, z, and the direction-cosines of its inward normal being l, m, n. If p is the pressure of the gas on this surface, the total of all the forces which this element of surface exerts on all the molecules of the gas has components

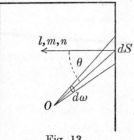

$$lp\,dS, \quad mp\,dS, \quad np\,dS.$$

Hence the contribution of all these forces to the virial is

$$-\tfrac{1}{2}(lx + my + nz)\,p\,dS. \quad \dots\dots(78)$$

Fig. 13

Let r be the distance of dS from the origin, and θ the angle between the normal to dS and the line joining the origin to dS (fig. 13). Then

$$lx + my + nz = -r\cos\theta,$$

and expression (78) may be replaced by $\tfrac{1}{2}r\cos\theta\,p\,dS$ or $\tfrac{1}{2}r^3 p\,d\omega$, where $d\omega$ is the element of solid angle which dS subtends at the origin.

If p is assumed to have the same value at all points of the surface, the total contribution to the virial made by all parts of the surface is

$$\tfrac{1}{2}p\iint r^3\,d\omega,$$

where the integral extends over the whole surface, and this is equal to

$$\tfrac{3}{2}pv,$$

where v is the whole volume of the containing vessel. Equation (77) now becomes

$$\Sigma\tfrac{1}{2}mc^2 = \tfrac{3}{2}pv - \tfrac{1}{2}\Sigma(xX + yY + zZ), \quad \dots\dots(79)$$

where only intermolecular forces are now to be included in the last term.

49. Suppose that the molecules are infinitesimal in size, but not necessarily spherical in shape, and that they exert no forces except **when in** actual contact. When a collision occurs, action and reaction are equal and opposite, so that X, Y, Z have equal and opposite values for the two molecules concerned. Since we are supposing the molecules to be infinitesimal in size, x, y, z have the same value for the two molecules, so that the contribution which all the intermolecular forces, including the forces of collision, make to

$$\Sigma(xX + yY + zZ)$$

is nil. Since we are supposing these to be the only forces in action, equation (79) becomes

$$pv = \Sigma \tfrac{1}{3}mc^2,$$

which is the equation already obtained for the pressure in an ideal gas.

Whatever the shape and size of the molecules, and whatever the forces between them may be, the corrections to be applied to this simple equation are all contained in the more general equation (79), which may be written in the form

$$pv = \Sigma \tfrac{1}{3}mc^2 + \tfrac{1}{3}\Sigma(xX + yY + zZ). \qquad \ldots \ldots (80)$$

50. This equation shews that the pressure may be regarded as made up of two contributions, one coming from the kinetic energy of motion of the molecules, and the other from the potential energy of their intermolecular forces, including of course the forces of collision.

The essence of the kinetic theory view of pressure is that it stresses this first contribution; the second figures only as a small correcting term. Newton had shewn* that if the pressure of a gas was proportional to its density, this pressure could be accounted for by repulsive forces between the particles of the gas; this view of course neglects the motion of the molecules and attributes the whole pressure to the second term in equation (80). We can, however, see that this view is untenable, for the following reason.

Let us imagine the containing vessel, together with the gas inside it, expanded to n times its original linear dimensions. The coordinates x, y, z are all increased n-fold, and as observation tells

---

* *Principia*, Prop. xxiii, theorem xviii.

us that pv remains constant, equation (80) shews that X, Y, Z must be decreased to $1/n$ times their original values. In other words, the intermolecular forces must vary inversely as the distance, as indeed Newton shewed. This is, however, an impossible law, since it would make the action of the distant parts of the gas preponderate over that of the contiguous parts, and so would not give a pressure which would be constant for a given volume and temperature as we passed from one vessel to another, or even from one part to another of the surface of the same vessel. We therefore conclude that the pressure of a gas cannot be explained by assuming repulsive forces between the molecules; it must arise mainly from the *motion* of the molecules.

51. We next proceed to calculate the contribution of the intermolecular forces to the virial. Let us assume that the force between two molecules is a repulsive force $\phi(r)$, which depends only on the distance r they are apart. If the centres of the molecules are at x, y, z, x', y', z', and if X, Y, Z, X', Y', Z' are the components of the forces acting on them, then

$$X = \phi(r)\frac{x-x'}{r}, \quad X' = \phi(r)\frac{x'-x}{r}, \text{ etc.,}$$

so that the contribution to ΣxX made by the force between these two particles is

$$xX + x'X' = \frac{\phi(r)}{r}(x-x')^2.$$

The contribution to $\Sigma(xX + yY + zZ)$ is therefore

$$\frac{\phi(r)}{r}\{(x-x')^2 + (y-y')^2 + (z-z')^2\} = r\phi(r).$$

Thus equation (80) may be replaced by

$$pv = \tfrac{1}{3}\Sigma mc^2 + \tfrac{1}{3}\Sigma\Sigma r\phi(r), \qquad \ldots\ldots(81)$$

where the summation extends over all pairs of molecules.

52. Since the gas contains N molecules, the summation on the right of equation (81) must be taken over $\tfrac{1}{2}N(N-1)$ pairs of molecules.

Let A, B be the two molecules of any such pair. If the molecules were simply points exerting no forces on one another, the chance of A and B being at a distance between r and $r + dr$ from one another would be

$$\frac{4\pi r^2 \, dr}{v}, \qquad\qquad \text{......(82)}$$

the numerator being the volume of the shell of thickness dr surrounding molecule A, and the denominator v being the whole space possible for the centre of B. Thus the number of pairs of molecules having their centres within a distance r of one another would be

$$\frac{2\pi N^2}{v} r^2 \, dr, \qquad\qquad \text{......(83)}$$

since it is obviously legitimate to neglect the difference between $N - 1$ and N.

Suppose, however, that molecules at a distance r repel one another with a force $\phi(r)$. Then we shall find, in § 228, that the probability of finding two molecules at a distance r apart is less than the probability of finding the same two molecules at a distance ∞ apart (∞ here denoting any distance great enough for the molecules to be out of range of each other's action) by a factor

$$e^{-2h\chi}, \qquad\qquad \text{......(84)}$$

where χ is the work which must be done in bringing the two molecules together from an infinite distance, so that

$$\chi = \int_r^\infty \phi(r) \, dr.$$

Making this correction to expression (83), the number of pairs of molecules at a distance r apart will be

$$\frac{2\pi N^2}{v} r^2 \, e^{-2h\chi} \, dr.$$

Multiplying this by $\frac{1}{3} r \phi(r)$ and integrating over all values of r,

$$\tfrac{1}{3}\Sigma\Sigma r\phi(r) = \frac{2\pi N^2}{3v} \int_0^\infty r^3 \phi(r) \, e^{-2h\chi} \, dr, \qquad \text{......(85)}$$

or, since

$$\phi(r) = -\frac{d\chi}{dr},$$

$$\tfrac{1}{3}\Sigma\Sigma r\phi(r) = \frac{2\pi N^2}{3v} \int_0^\infty r^3 \frac{d\chi}{dr} e^{-2h\chi} \, dr.$$

Replacing $\frac{1}{3}\Sigma mc^2$ by its value NRT, equation (81) assumes the form

$$pv = NRT + \frac{2\pi N^2}{3v}\int_0^\infty r^3 \frac{d\chi}{dr} e^{-2h\chi}\, dr$$

or

$$pv = NRT\left(1 + \frac{B}{v}\right), \qquad \ldots\ldots(86)$$

where

$$B = \frac{2\pi N}{3RT}\int_0^\infty r^3 \frac{d\chi}{dr} e^{-2h\chi}\, dr. \qquad \ldots\ldots(87)$$

Replacing $1/RT$ by its value $2h$, and integrating by parts, we readily find that B may be put in the alternative form

$$B = 2\pi N\int_0^\infty r^2(1 - e^{-2h\chi})\, dr. \qquad \ldots\ldots(88)$$

53. If the molecules approximate closely to hard elastic spheres, the forces between molecules only come into play when r is very nearly equal to σ; for all other values of r,

$$\frac{d\chi}{dr} = 0.$$

Thus we may replace r^3 by σ^3 in equation (87) and find

$$B = \frac{2\pi N\sigma^3}{3RT}\int_0^\infty e^{-2h\chi}\frac{d\chi}{dr}\, dr$$

$$= \frac{\pi N\sigma^3}{3hRT} = \tfrac{2}{3}\pi N\sigma^3 = b,$$

where b is the b of Van der Waals' equation.

Equation (86) now becomes

$$pv = NRT\left(1 + \frac{b}{v}\right), \qquad \ldots\ldots(89)$$

which agrees with Van der Waals' equation (72).

Clearly, then, equation (86) is a generalisation of Van der Waals' equation, the general quantity B, defined by equation (88), replacing the simple b of Van der Waals. The quantity B is generally known as the Second Virial Coefficient.

If forces of cohesion of the kind specified in §45 are also supposed to act, these forces will have a contribution to make to the virial. To the first order of small quantities, we may, in calculating $\Sigma\Sigma r\phi(r)$, ignore the effect of the forces of cohesion on the distribution of density of the gas. The value of $\Sigma\Sigma r\phi(r)$ is therefore obviously proportional simply to ρ^2 per unit volume of the gas. Allowing for this addition to the virial, equation (89) must be replaced by

$$pv = NRT\left(1+\frac{B}{v}\right) - c\rho^2 v,$$

where c is the same as the c of § 45, and is independent of the temperature. Or again this last equation may be written

$$\left(p+\frac{a}{v^2}\right)v = NRT\left(1+\frac{B}{v}\right),$$

agreeing with Van der Waals' equation (73) as far as the first order of small quantities.

54. The quantity B is easily evaluated when the repulsive force $\phi(r)$ varies as an inverse power of the distance, say $\phi(r) = \mu r^{-s}$. We then have

$$\chi = \int_r^\infty \phi(r)\,dr = \frac{1}{s-1}\frac{\mu}{r^{s-1}},$$

so that
$$\Sigma\Sigma r\phi(r) = \frac{2\pi N^2}{v}\int_0^\infty \frac{\mu}{r^{s-3}} e^{-\left(\frac{2h}{s-1}\frac{\mu}{r^{s-1}}\right)}\,dr$$

$$= \frac{2\pi N^2}{2hv}\left(\frac{2h\mu}{s-1}\right)^{\frac{3}{s-1}} \Gamma\left(1-\frac{3}{s-1}\right).$$

It follows that B can be put in the form

$$B = \tfrac{2}{3}\pi N\sigma^3,$$

where σ^3 is given by

$$\sigma^3 = \left(\frac{2h\mu}{s-1}\right)^{\frac{3}{s-1}} \Gamma\left(1-\frac{3}{s-1}\right). \qquad \ldots\ldots(90)$$

This amounts to supposing that the molecules behave like elastic spheres, but that these have a diameter σ which depends on h, and so varies with the temperature; this merely represents the fact that at high temperatures the collisions are more violent, so that the molecules penetrate farther into each other's fields of force before being brought to relative rest.

If B_0 and σ_0 are the values of B and σ at $0°$ C., the general values at temperature T are given by

$$B = B_0\left(\frac{T}{273 \cdot 2}\right)^{-\frac{3}{s-1}},$$

$$\sigma = \sigma_0\left(\frac{T}{273 \cdot 2}\right)^{-\frac{1}{s-1}}.$$

55. The evaluation of B for more complicated laws of force presents a very intricate problem of mathematics. Keesom[*] has studied the problem when the molecules are regarded as rigid spheres of diameter σ surrounded by an attractive field of force proportional to μr^{-s}, and has evaluated B in the form of an infinite series

$$B = \tfrac{2}{3}\pi N \sigma^3\left[1 - 3\sum_{n=1}^{n=\infty}\frac{(2hu)^n}{n!\,\{(s-1)\,n-3\}}\right],$$

where u is the work done in separating two molecules which are in contact, and removing them to an infinite distance apart. Keesom has also[†] evaluated B when the molecules are rigid spheres each containing an electric doublet at its centre, the value again being expressed as an infinite series.

J. E. Lennard-Jones[‡] has discussed the problem when the molecules are supposed to repel according to the law of force[§]

$$\frac{\lambda_n}{r^n} - \frac{\lambda_m}{r^m},$$

and has obtained for B the value

$$B = \tfrac{2}{3}\pi N\left(\frac{\lambda_n}{\lambda_m}\frac{m-1}{n-1}\right)^{\frac{3}{n-m}} F(y), \qquad \ldots\ldots(91)$$

where
$$y = \frac{2h\lambda_m}{m-1}\left(\frac{n-1}{2h\lambda_n}\right)^{\frac{m-1}{n-1}}$$

and $\quad F(y) = y^{\frac{3}{n-m}}\left\{\Gamma\left(\frac{n-4}{n-1}\right) - \sum_{\tau=1}^{\tau=\infty}\frac{3\Gamma\left(\dfrac{\tau(m-1)-3}{n-1}\right)}{\tau!\,(n-1)}y^\tau\right\}.$

This contains the two simple formulae previously given as special cases.

* *Comm. Phys. Lab. Leiden*, Supp. 24 B (1912), p. 32, and *Proc. Amsterdam Akad.* 15 (1912), p. 1406.　　　　† *L.c. ante.*
‡ *Proc. Roy. Soc.* 106 A (1924), p. 463.　　　　§ See § 149 below.

GAS THERMOMETRY

56. In spite of its imperfections, the equation of Van der Waals undoubtedly provides the most convenient basis for discussing the behaviour of a gas over those ranges of pressure, density and temperature within which the deviation from Boyle's law is small. We consider now some of the physical properties of a gas predicted by the equation of Van der Waals.

Changes of Constant Volume

57. Let us first suppose the volume of the gas to be kept constant, and the temperature be raised from T_0 to T_1.

If p_0, p_1 are the pressures corresponding to these temperatures, Van der Waals' equation tells us that

$$\left(p_0 + \frac{a}{v^2}\right)(v-b) = NRT_0, \qquad \text{......(92)}$$

$$\left(p_1 + \frac{a}{v^2}\right)(v-b) = NRT_1. \qquad \text{......(93)}$$

By subtraction we get

$$(p_1 - p_0)(v-b) = NR(T_1 - T_0).$$

This shews that the increase in pressure is proportional to the increase in temperature; in fact p_1 is given by the formula

$$p_1 = p_0\{1 + \kappa_p(T_1 - T_0)\},$$

where κ_p is a "pressure-coefficient" given by

$$\kappa_p = \frac{NR}{p_0(v-b)}, \qquad \text{......(94)}$$

or, from equation (92), by

$$\kappa_p = \left(1 + \frac{a}{p_0 v^2}\right)\frac{1}{T_0}. \qquad \text{......(95)}$$

In practical work, the initial temperature T_0 is usually taken to be $0°$ C., so that if θ is the temperature on the centigrade scale

$$p_1 = p_0(1 + \kappa_p \theta).$$

Equation (95) shews that κ_p depends on the density but not on the temperature, so that *for a given density of gas, the pressure-coefficient is independent of the temperature.*

This law naturally is true only within the limits in which Van der Waals' equation is true. Regnault* proved that it was very approximately true under ordinary conditions by filling gas thermometers with different gases and shewing that they gave identical readings over a large range of temperature. More recent and more exact experiments have shewn, as might be expected, that the law is by no means absolutely exact or of universal validity. Full tables of values of κ_p will be found in the *Recueil de constantes physiques.*† As a specimen may be given the following values, obtained by Chappuis in 1903.‡

Values of κ_p

Temperature	For nitrogen ($p_0 = 1001 \cdot 9$ mm. at 0° C.)	For CO_2 ($p_0 = 998 \cdot 5$ mm. at 0° C.)
0° to 20°	$\kappa_p = 0 \cdot 0036754$	$\kappa_p = 0 \cdot 0037335$
0° to 40°	$0 \cdot 0036752$	$0 \cdot 0037299$
0° to 100°	$0 \cdot 0036744$	$0 \cdot 0037262$

Callendar§ gives the following values for the pressure-coefficients (0° to 100° C. at initial pressure 1000 mm.) of three of the more permanent gases:

$$\text{Air} \qquad 0 \cdot 00367425,$$
$$\text{Nitrogen} \qquad 0 \cdot 00367466,$$
$$\text{Hydrogen} \qquad 0 \cdot 00366254,$$

while for neon and argon, Leduc¶ has found the values:

Neon $T_0 = 5 \cdot 47°$ C. to $T_1 = 29 \cdot 07°$ C., $\kappa_p = 0 \cdot 003664$,

Argon $T_0 = 11 \cdot 95°$ C. to $T_1 = 31 \cdot 87°$ C., $\kappa_p = 0 \cdot 003669$.

A gas in which the pressure obeys the Boyle-Charles law $pv = NRT$ exactly is described as a "perfect" gas. For such a

---

* *Mém. de l'Acad.* 21, p. 180.

† Pp. 234–40. The pressure-coefficient κ_p from range θ to θ' is there denoted by $\beta^{\theta'}$. ‡ *L.c.* p. 234.

§ *Phil. Mag.* 5, p. 92. ¶ *Comptes Rendus,* 164 (1917), p. 1003.

gas, $a = 0$, so that $\kappa_p = \dfrac{1}{273 \cdot 2}$ or $0 \cdot 003660$. It will be noticed that the pressure-coefficient of hydrogen approximates very nearly to that of a perfect gas, shewing that the value of a is extremely small for hydrogen. Thus the value of κ_p is almost independent of the volume of the gas. For this reason the *Comité internationale des poids et mesures* decided on the constant volume hydrogen thermometer as standard gas-thermometer. The small value which is known to exist for a is recognised in the stipulation of the committee that the volume at which the gas is used is to be such that there is a pressure of 1000 mm. at 0° C. But nitrogen gas-thermometers are also employed, and are found to possess certain advantages.

Changes at Constant Pressure

58. Let us next suppose that the pressure is kept at the constant value p, and that changing the temperature from T_0 to T_1 is found to change the volume from v_0 to v_1. The two equations analogous to (92) and (93) of § 57 are

$$\left(p + \frac{a}{v_0^2}\right)(v_0 - b) = NRT_0, \qquad \ldots\ldots(96)$$

$$\left(p + \frac{a}{v_1^2}\right)(v_1 - b) = NRT_1. \qquad \ldots\ldots(97)$$

Neglecting the product ab which is a small quantity of the second order, we obtain, on subtraction,

$$\left(p - \frac{a}{v_0 v_1}\right)(v_1 - v_0) = NR(T_1 - T_0). \qquad \ldots\ldots(98)$$

We find from this that v_1 is given by

$$v_1 = v_0\{1 + \kappa_v(T_1 - T_0)\},$$

where κ_v is a "volume-coefficient" given by

$$\kappa_v = \frac{NR}{pv_0 - a/v_1}.$$

On eliminating NR between this and equation (96), we obtain

$$\kappa_v = \left\{1 + \frac{a}{pv_0}\left(\frac{1}{v_0} + \frac{1}{v_1}\right) - \frac{b}{v_0}\right\}\frac{1}{T_0}. \qquad \ldots\ldots(99)$$

This is more complicated than the formula for the pressure-coefficient (95) in that it depends both on the volume and the pressure.

For a perfect gas, $a = b = 0$, and the value of κ_v, like that of κ_p, becomes $0 \cdot 003660$. The following table, similar to that given on p. 79, contains some observed values of κ_v.

Values of κ_v

Temp.	For nitrogen ($p = 1001 \cdot 9$ mm.)	For CO_2 ($p = 998 \cdot 5$ mm.)	For CO_2 ($p = 517 \cdot 9$ mm.)
0° to 20° 0° to 40° 0° to 100°	$\kappa_v = 0 \cdot 0036770$ $0 \cdot 0036750$ $0 \cdot 0036732$	$\kappa_v = 0 \cdot 0037603$ $0 \cdot 0037536$ $0 \cdot 0037410$	$\kappa_v = 0 \cdot 0037128$ $0 \cdot 0037100$ $0 \cdot 0037073$

Further values will be found in the *Recueil de constantes physiques*, from which the above are taken.

The absolute scale of temperature can, as we have seen, be defined without reference to any particular substance. This makes it a perfect scale for abstract problems and theoretical discussions, but no thermometer can ever be constructed to read absolute temperatures directly. If any gas existed which was "perfect" in the sense explained above, a gas-thermometer using this gas would give readings of absolute temperature. In the absence of such a perfect gas, the absolute scale can be approximately realised by making the necessary small corrections to gas-thermometers in which ordinary gases are used. These corrections depend on the extent to which the values of κ_p and κ_v for the gases used differ from the theoretical value $0 \cdot 003660$ for a perfect gas. Tables of such corrections will be found in books of physical constants.*

Evaluation of a and b

59. From an experimental evaluation of the "pressure-coefficient" κ_p given by equation (95), the quantity a can be obtained at once, and when this is known, the value of b can be obtained from the observed value of the volume-coefficient.

---

* E.g. Kaye and Laby, *Physical Constants*, 8th edn (1936), p. 47.

For instance, using Callendar's value for κ_p for air at a pressure of 1000 mm. mercury, we have (equation (95)), with $T_0 = 273 \cdot 2$,

$$\kappa_p = \left(1 + \frac{a}{p_0 v^2}\right) \frac{1}{T_0} = 0 \cdot 003673,$$

while
$$\frac{1}{T_0} = 0 \cdot 003660.$$

At this pressure, therefore,

$$\frac{a}{v^2} = p_0 T_0 \times 0 \cdot 000013 = 0 \cdot 0047 \text{ atmosphere pressure.}$$

Thus for air at $1 \cdot 3158$ atmospheres pressure at the boundary, the forces of cohesion result in an apparent diminution of pressure of $0 \cdot 0047$ atmosphere, or about one-three hundredth of the whole, so that the pressure in the interior of the gas is $1 \cdot 3205$ atmospheres. This gives some idea of the magnitude of the forces of cohesion.

At other pressures, a/v^2 is proportional to $1/v^2$, and so to the square of the pressure. For instance, at a pressure of 1 atmosphere,

$$\frac{a}{v^2} = 2649 \cdot 5 \text{ in C.G.S. units} = 0 \cdot 00260 \text{ atmosphere.}$$

When a has been determined in this way, we can determine b from the observed values of κ_v. This determination is of special interest, because from it we can calculate directly the value of σ, the diameter of the molecule or of its sphere of molecular action. From the discussion of a great number of experiments by Regnault, Van der Waals deduced the following values for b:

Air	$0 \cdot 0026$,
Carbon-dioxide	$0 \cdot 0030$,
Hydrogen	$0 \cdot 00069$.

These values refer to a mass of gas which occupies unit volume at a pressure of 1000 mm. of mercury.

A more recent method of determining b depends on the measurement of the Joule-Thomson effect. Calculations by Rose-Innes[*] lead to the following values for b:

Air $1 \cdot 62$,	Nitrogen $2 \cdot 03$,	Hydrogen $10 \cdot 73$.

[*] *Phil. Mag.* **2** (1901), p. 130.

Referred to a cubic centimetre of gas at normal pressure,

Air	0·00209,
Nitrogen	0·00255,
Hydrogen	0·00096.

For helium, Kamerlingh Onnes* has determined the value for b:

Helium 0·000432.

Values of Molecular radius $\frac{1}{2}\sigma$

60. The value of b is, as in equation (70), equal to $\frac{2}{3}N\pi\sigma^3$, and since the values of b have been determined for a cubic cm. of gas at normal pressure, we may take $N = 2\cdot69 \times 10^{19}$ (§ 15), and so determine σ immediately.

The values of $\frac{1}{2}\sigma$ deduced from the best values of b are as follows:

Gas	Value of b (1 c.c. of gas)	Observer	Value of $\frac{1}{2}\sigma$
Hydrogen	0·00096	Rose-Innes	$1\cdot27 \times 10^{-8}$
Helium	0·000432	Kamerlingh Onnes	$0\cdot99 \times 10^{-8}$
Nitrogen	0·00255	Rose-Innes	$1\cdot78 \times 10^{-8}$
Air	0·00209	Rose-Innes	$1\cdot66 \times 10^{-8}$
Carbon-dioxide	0·00228	Van der Waals	$1\cdot71 \times 10^{-8}$

ISOTHERMALS

61. One of the most instructive ways of representing the relation between the pressure, volume and temperature of a gas is by drawing "isothermals" or graphs shewing the relation between pressure and volume when the temperature is kept constant. There will of course be one isothermal corresponding to every possible temperature, and if all the isothermals are imagined drawn on a diagram in which the ordinates and abscissae represent pressure and volume respectively, we shall have a complete representation of the relation in question.

* *Comm. Phys. Lab. Leiden,* **102**a (1907), p. 8.

Isothermals of an Ideal Gas

62. For an ideal gas, the relation is expressed by the equation

$$pv = NRT. \qquad\qquad ......(100)$$

To represent this relation by means of isothermals, we take p and v as rectangular axes and draw the group of curves obtained by assigning different constant values to T in equation (100). The curves all have equations of the form $pv = $ const., and so are a system of rectangular hyperbolas, lying as in fig. 14. These are the isothermals of an ideal gas.

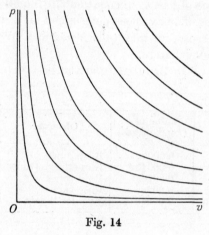

Fig. 14

Isothermals of a Gas obeying Van der Waals' Equation

63. Let us next examine what isothermals will correspond to Van der Waals' equation

$$\left(p + \frac{a}{v^2}\right)(v - b) = NRT. \qquad\qquad ......(101)$$

It will be noticed that if the system of curves shewn in fig. 14 is pushed bodily through a distance b parallel to the axis of v, they will give the system of isothermals represented by the equation

$$p(v - b) = NRT, \qquad\qquad ......(102)$$

and on further drawing down every ordinate through a distance

a/v^2 parallel to the axis of p, we shall obtain the system of isothermals represented by Van der Waals' equation (101).

The isothermals are found to lie as in fig. 15 in which the thick
line AB is the curve $p = -a/v^2$, while the line BCD is $v = b$.

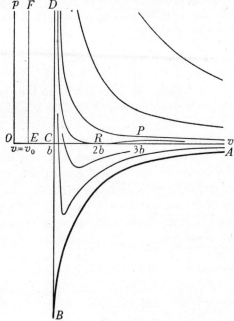

Fig. 15

The isothermals corresponding to high temperatures naturally
lie exactly like those in the earlier fig. 14, the isothermals of an
ideal gas, but at lower temperatures divergences begin to appear.
Below a certain temperature, the value of p does not steadily increase as v decreases; on the contrary, after increasing for a time,
the value of p reaches a maximum, then decreases to a minimum,
after which it again increases.

These maxima and minima must of course occur at the points
at which $dp/dv = 0$.

From equation (101), we may write the equation of the isothermals in the form

$$\log\left(p + \frac{a}{v^2}\right) + \log(v - b) = \text{const.,} \qquad \ldots\ldots(103)$$

and the points at which $dp/dv = 0$ are seen to be given by

$$p = \frac{a(v - 2b)}{v^3}.$$

Clearly this value of p is positive for all values of v between $2b$ and ∞. Within this range it attains a maximum value, which is readily found to be given by

$$v = 3b, \quad p = \frac{a}{27b^2}, \quad NRT = \frac{8a}{27b}. \qquad \ldots\ldots(104)$$

The isothermals for values of T greater than that given by this equation can have no points at which $dp/dv = 0$, and so are everywhere convex to the axis of v.

64. The isothermals of a real gas will lie like those shewn in fig. 15 so long only as the gas does not differ too much from an ideal gas. The curves in fig. 15 will accordingly represent the isothermals of a real gas with accuracy in the regions far removed from both axes, but not near to these axes. We must inquire what alterations must be made in these curves in order to represent the isothermals of a real gas.

The isothermal $T = 0$ is represented in fig. 15 by the broken line made up of the curve AB and the vertical line BCD. The true isothermal is, however, known with accuracy. If a gas is cooled to temperature $T = 0$ and is then compressed, the pressure remains zero until the molecules are actually in contact. Let the volume in this state be denoted by v_0. The pressure may now be increased to any extent and the volume will still retain the same value v_0, this being the smallest volume which can be occupied by the molecules. Now v_0, being the smallest volume into which N spheres each of diameter σ can be compressed, is easily found to be given by

$$v_0 = \frac{N\sigma^3}{\sqrt{2}},$$

while the value of b is (cf. equation (70))

$$b = \tfrac{2}{3}N\pi\sigma^3 = 2 \cdot 96 v_0. \qquad \ldots\ldots(105)$$

Thus the true isothermal $T = 0$, instead of being the curve $ABCD$ in fig. 15, consists of the two lines vE, EF. If we imagine the curves in fig. 15 so distorted that the point B is made to coincide with the point E, and the curve $ABCD$ with the lines vEF, we shall obtain an idea of the run of the isothermals of a real gas. The curves will perhaps lie somewhat as in fig. 16, in which both the vertical and horizontal scales have been largely increased over those employed in fig. 15, but the vertical scale has been increased much more than the horizontal.

Isothermals of a gas obeying Dieterici's Equation

65. The isothermals of a gas which is supposed to obey Dieterici's equation may be discussed in a similar manner.

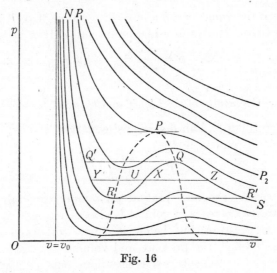

Fig. 16

If we neglect the exponential factor, the isothermals are again given by equation (102), namely

$$p(v-b) = NRT,$$

and, as before, the curves may be obtained by shifting the isothermals of an ideal gas (fig. 14) through a distance b parallel to the axis of v. If we now restore the exponential factor, we must reduce the value of p at each point of each isothermal by a factor $e^{-a/NRTv}$, and the resulting curves will be the isothermals required. They will again be found to lie as in fig. 16.

Again the network of isothermals for the lower temperatures contain regions in which p and v are decreasing together. The equation of an isothermal may be written in the form

$$\log p + \log (v-b) + \frac{a}{NRTv} = \text{const.},$$

so that the points at which p has its maximum and minimum values on any isothermal are given by

$$\frac{dp}{dv} = 0 \quad \text{or} \quad \frac{1}{v-b} = \frac{a}{NRTv^2},$$

so that
$$NRT = \frac{a(v-b)}{v^2}.$$

Between the values $v = b$ and $v = \infty$, we see that T is everywhere positive. It attains a maximum value given by

$$v = 2b, \quad p = \frac{a}{4b^2}e^{-2}, \quad NRT = \frac{a}{4b}. \qquad \ldots\ldots(106)$$

Continuity of the Liquid and Gaseous States

66. According to Van der Waals' equation, the temperature determined by equation (104),

$$T = \frac{8a}{27NRb},$$

has the special property that at all temperatures above this temperature a decrease of volume is always accompanied by an increase of pressure, but below it every isothermal contains stretches in which the pressure and volume decrease together. Every point on these latter stretches represents a collapsible or unstable state, since, if the gas is already yielding to pressure, each contraction increases the disparity between the resistance of the gas and the force exerted on it—the gas is like an army in retreat which becomes more and more demoralised with every yard it retreats.

The equation of Dieterici makes exactly similar predictions except that the critical temperature, determined by equation (106), is given by

$$T = \frac{a}{4NRb}.$$

Let us fix our attention on any one collapsible state of the kind predicted by both equations, say that represented by the point X in fig. 16. On the same isothermal as X, there must clearly be two other points Y, Z which represent states having the same temperature and pressure as X. At each of these two points dp/dv is positive, so that the two states in question are both stable; they ought therefore both to be known to observation. The point Z obviously represents the gaseous state; *the point Y corresponding to lesser volume is believed to represent the liquid state.*

With this interpretation it is at once clear that so long as a gas is kept at a temperature above that of the isothermal $P_1 P P_2$, no amount of compression can force it into the liquid state. The temperature of the isothermal $P_1 P P_2$ is called the "critical temperature" of the substance. We see that *so long as a gas is kept above the critical temperature, no pressure, however great, can liquefy it.*

67. A gas which is below the critical temperature is usually described as a vapour. We therefore see that the line PP_2 in fig. 16 is the line of demarcation between the gaseous and vapour states, while PP_1 is the line of demarcation between the gaseous and liquid states.

It remains to examine the demarcation between the liquid and vapour states, which is at present represented by the unstable region in which dp/dv is positive. If U is any point in this region, common experience tells us that there is a *stable* state in which the pressure and volume are those of the point U. What is this state?

Let us imagine a line drawn through U parallel to the axis of v. Let this cut any isothermal in the points X, Y, Z, the two latter representing stable states—liquid and vapour respectively. As these two states have the same pressure, it follows that a quantity of vapour in the state Z can rest in equilibrium with a quantity of liquid in state Y. By choosing these quantities in a suitable ratio, the combination of the two will be represented by the point U. Here, then, we have an interpretation of the physical meaning of the point U. As the vapour is compressed at the temperature of the isothermal $SZQXRY$, the substance remains a vapour until the point Z is reached. At this point condensation sets in; some of the substance is in the state Z but some also is in the state Y. As the ratio of these amounts changes, the representative point moves along the straight line $ZXUY$ until, by the time the point Y is reached, the whole of the matter is in the liquid state. After this the substance, remaining wholly in the liquid state, moves through the series of changes represented by the path $YQ'N$.

There is of course an element of arbitrariness in this, for instead of describing the path $SZUYN$ the substance might equally well have been supposed to describe the path $SR'RYN$, keeping at the same temperature throughout; or any other path composed of two stable branches of an isothermal joined by a line of constant pressure. In other words, there is no unique relation between the pressure and temperature of evaporation or condensation. This is, however, in accordance with the known properties of matter; there are such things in nature as super-cooled vapours which may be represented by the range ZQ in fig. 16, and super-heated liquids, which may be represented on the range YR.

Under normal conditions, however, when there are no complications produced by surface-tensions, particles of dust, or other impurities, there must be a definite boiling point corresponding to every pressure, and the path of a substance from one state to another, given the same external conditions, must be quite definite. So far we have not arrived at any such definiteness.

Maxwell[*] and Clausius[†] both attempted to obtain definite paths for a substance changing at a constant temperature. They reached the conclusion that if the line $SZXYN$ in fig. 16 is to represent the actual isothermal path from S to N, it must be so chosen that the areas ZQX, XRY are equal. For, imagine the substance starting from Z, and passing through the cycle of changes represented in fig. 16 by the path $ZQXRYXZ$, the first part of the path $ZQXRY$ being along the curved isothermal, and the second part YXZ along the straight line. Since this is a closed cycle of changes, it follows from the second law of thermodynamics that $\int \frac{dQ}{T} = 0$, where dQ is the total heat supplied to

* Collected Works, 2, p. 425.   † Wied. Ann. 9 (1880), p. 337.

the substance in any small part of the cycle and the integral is taken round the whole closed path representing the cycle. Since the temperature is constant throughout the motion, this equation becomes $\int dQ = 0$, so that the integral work done on the gas throughout the cycle is *nil*. As in § 12, this work is equal to $\int p\,dv$ and therefore to the area, measured algebraically, of the curve in fig. 16 which represents the cycle. Hence this area must vanish, which is the result already stated.

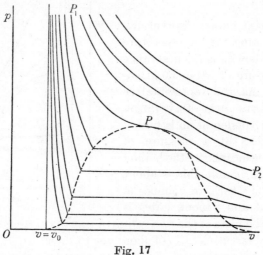

Fig. 17

68. The figure which is obtained from fig. 16 upon replacing the curved parts of isothermals such as $ZQXRY$ by the straight line ZXY is represented in fig. 17. This figure ought accordingly to represent the main features of the observed systems of isothermals of actual substances.

Comparison with Experiment

69. Fig. 18 shows the isothermals of crabon-dioxide as found in the classical experiments of Andrews.* The figures on the left hand denote pressure measured in atmospheres (the isothermals

* *Phil. Trans.* **159** (1869), p. 575 and **167** (1876), p. 421.

only being shewn for pressures above 47 atmospheres), while those on the right denote the temperatures centigrade of the various isothermals.

The isothermal corresponding to the temperature 31·1° is of great interest, as being very near to the critical isothermal, the value of the critical temperature being given by Andrews as 30·92° and by Keesom as 30·98°.

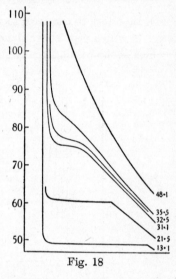

On this isothermal, as on all those above it, the substance remains gaseous, no matter how great the pressure.

The next lower isothermal, corresponding to temperature 21·5° C., shows a horizontal range at a pressure of about 60 atmospheres. As the representative point moves over this range, boiling or condensation is taking place. Thus at a pressure of about 60 atmospheres the boiling point of carbon-dioxide is about 21·5° C. The ratio of volumes in the liquid and vapour states is

Fig. 18

equal to the ratio of the two values of v at the extremities of the horizontal range—a ratio of about one to three.

The lowest isothermal of all corresponds to a temperature of 13·1° C. Here the inequality between the volumes of the liquid and the gas is greater than before. In fact an examination of the general theoretical diagram given in fig. 17 shews that as the temperature decreases the inequality must become more and more marked, so that in all substances the distinction between the liquid and gaseous states becomes continually more pronounced as we recede from the critical temperature.

The Critical Point

70. The point represented by P in fig. 16 is generally described as the "critical point". Just as there are points in England at which three counties meet, so here is a point where three states meet—the liquid, the vapour and the gaseous. And, because of this, a substance at the critical point possesses certain remarkable properties.

The values of the temperature, pressure and volume at the critical point are usually denoted by T_c, p_c and v_c. The temperature T_c is of course the critical temperature, above which liquefaction is impossible; p_c and v_c may be defined as the pressure and volume at which liquefaction first begins when the substance is at a temperature just below T_c.

We have seen that the existence of this critical point is implied in the equations of both Van der Waals and Dieterici. The two equations make different predictions as to the position of the point, but this is not surprising, since both equations are only true when the deviations from Boyle's law are small, and the critical point is in a region in which these deviations are quite definitely not small. Thus both predictions as to the position of the critical point are unwarranted extrapolations. The predictions are shewn in the following table:

Equation of	NRT_c	p_c	v_c	$\dfrac{NRT_c}{p_c v_c}$
Van der Waals	$0.30\,\dfrac{a}{b}$	$0.037\,\dfrac{a}{b^2}$	$3.00b$	2.67
Dieterici	$0.25\,\dfrac{a}{b}$	$0.034\,\dfrac{a}{b^2}$	$2.00b$	3.69

The last column is of interest from the fact that $NRT_c/p_c v_c$ is, on any theory, a pure number, having no physical dimensions at all. If Boyle's law still held at the critical point, this number would of course be equal to unity; we see how far the critical point is from the regions in which Boyle's law holds.

Actual observations of critical data are recorded in the table on p. 94 overleaf.

These data shew that the properties of different gases vary widely, and that for most gases neither the equation of Van der Waals nor that of Dieterici comes particularly near to the truth. Generally speaking, however, the equation of Dieterici appears to be more accurate than that of Van der Waals, particularly for the heavier and more complex gases.

Gas	$\dfrac{v_c}{b}$	$\dfrac{NRT_c}{p_c v_c}$
Hydrogen	2·80	3·27
Helium	—	3·26
Nitrogen	1·50	3·42
Oxygen	1·46	3·42
Neon	—	3·42
Argon	—	3·42
Xenon	—	3·60
CO_2	1·86	3·61
HCl	—	3·48
CCl_4	—	3·68
C_6H_6	—	3·71
C_6H_5Br	—	3·78
C_6H_5Cl	—	3·81
Van der Waals	3·00	2·67
Dieterici	2·00	3·69

71. It is interesting to compare the two equations at places other than the critical point.

According to the equation of Dieterici,

$$e^{a/NRTv} = \frac{NRT}{pv(1-b/v)},$$

so that, on taking logarithms of both sides,

$$\frac{a}{NRTv} = \log\left(\frac{NRT}{pv}\right) + \frac{b}{v} + \frac{1}{2}\left(\frac{b}{v}\right)^2 + \frac{1}{3}\left(\frac{b}{v}\right)^3 + \dots$$

and

$$\frac{a}{NRT} = v\log\left(\frac{NRT}{pv}\right) + b + \frac{b^2}{2v} + \frac{b^3}{3v^2} + \dots.$$

According to Van der Waals,

$$p(v-b) = NRT - \frac{a}{v^2}(v-b),$$

whence we obtain

$$\frac{a}{NRT} = v\log\left(\frac{NRT}{pv}\right) + b + \frac{b^2}{2v} + \frac{b^3}{3v^2} + \dots$$

$$+ \frac{a}{NRT}\frac{b}{v} + \frac{1}{2}\left(\frac{a}{NRT}\right)^2\left(1 - \frac{b}{v}\right)^2 v + \dots.$$

If we neglect powers and products of the small quantities a and b, both equations predict that

$$v \log \left(\frac{NRT}{pv} \right)$$

should be constant along any isothermal. The following table gives, in its last column, the value of this quantity along the critical isothermal ($T = 187 \cdot 8°$ C.) of isopentane, as calculated from the observations of Young. The values of p in the third column are those given by Dieterici's equation (74) for the temperature $187 \cdot 8°$ C. and the pressure recorded in the second column.*

Critical Isothermal of Isopentane

Critical temperature $= 187 \cdot 8°$ C. Critical volume $= 4 \cdot 266$ c.c. per gramme.

Volume v, per gramme	Pressure p, mm. of mercury	Pressure p (calc.)	$v \log \left(\dfrac{NRT}{pv} \right)$
2·4	49080	42730	1·271
2·5	40560	35810	1·486
2·6	34980	32090	1·669
2·8	28940	28390	1·938
3·0	26460	26780	2·103
3·2	25490	26000	2·205
3·6	25050	25420	2·326
4·0	25020	25320	2·402
4·3	25010	25300	2·447
4·6	25000	25300	2·483
5	24990	25240	2·520
6	24840	24880	2·564
7	24400	—	2·577
8	23710	23400	2·582
9	22930	—	2·576
10	22040	21590	2·575
12	20300	19850	2·568
15	17980	17540	2·548
20	14840	14560	2·564
30	10950	10770	2·526
40	8570	8508	2·624
50	7068	7025	2·625
60	6001	5978	2·652
80	4614	4604	2·680
90	4132	4127	2·637
100	3750	3740	2·680

* *Ann. d. Phys.* **5** (1901), p. 58.

It appears that $v \log \left(\dfrac{NRT}{pv} \right)$ is approximately constant for all pressures less than about 12 atmospheres, shewing the range within which a first approximation holds. There is, however, very tolerably good agreement between the observed and calculated values of p far beyond this range, indicating considerable accuracy in Dieterici's equation.

Reduced Equation of State

72. The critical temperature, pressure and volume form noteworthy milestones in the various states of a gas, and so provide valuable units for the measurement of temperature, pressure and volume in general.

Let us introduce quantities t, $\mathfrak{p}$ and $\mathfrak{v}$ defined by

$$t = \frac{T}{T_c}, \quad \mathfrak{p} = \frac{p}{p_c}, \quad \mathfrak{v} = \frac{v}{v_c}, \qquad \ldots \ldots (107)$$

so that t denotes the ratio of the temperature of any substance to its critical temperature, and so on. The quantities t, $\mathfrak{p}$, $\mathfrak{v}$ are called the reduced temperature, pressure and volume respectively.

According to Van der Waals' equation, we have (cf. equations (104))

$$p = \frac{a}{27b^2}\mathfrak{p}, \quad v = 3b\mathfrak{v}, \quad NRT = \frac{8a}{27b}t,$$

so that the equation reduces to

$$\left(\mathfrak{p} + \frac{3}{\mathfrak{v}^2} \right)\left(\mathfrak{v} - \frac{1}{3} \right) = \frac{8}{3}t. \qquad \ldots \ldots (108)$$

It will be noticed that this equation is the same for all gases, since the quantities a, b which vary from one gas to another have entirely disappeared. An equation, such as that of Van der Waals, which aims at expressing the relation between pressure, volume and temperature in a gas, is called an equation of state, or sometimes a characteristic equation or gas-equation. Equation (108) may be called the "reduced" equation of state of Van der Waals, and is the same for all gases.

On the other hand, according to Dieterici's equation, the values of t, $\mathfrak{p}$ and $\mathfrak{v}$ are given by equations (106) and on substituting these values for T_c, p_c, v_c in Dieterici's equation, we find that the corresponding "reduced" equation of state is

$$\mathfrak{p}(2\mathfrak{v} - 1) = \mathfrak{t}e^{2\left(1 - \frac{1}{\mathfrak{v}}\right)}. \qquad \ldots\ldots(109)$$

Corresponding States

73. If either of these equations could be regarded as absolutely true for all gases, it appears that when any two of the quantities $\mathfrak{t}$, $\mathfrak{p}$, $\mathfrak{v}$ are known, the third would also be given, and would be the same for all gases. In other words, there would be a relation of the form $\mathfrak{p} = f(\mathfrak{t}, \mathfrak{v})$, in which the coefficients in f would be independent of the nature of the gas.

The same is true for any equation of state whatever, provided that this contains only *two* quantities which depend on the particular structure of the gas in question—say, for instance, the same two as in the equation of Van der Waals, representing the size of the molecules and the cohesion-factor. For if a and b represent these two constants, the equation of state will be a relation between the five quantities

$$p, v, T, a \text{ and } b.$$

The critical point is fixed by two more equations, these being in fact $\dfrac{dp}{dv} = 0$ and $\dfrac{d^2p}{dv^2} = 0$, and these involve the critical quantities T_c, p_c and v_c as well as a and b. From the three equations just mentioned we can eliminate a and b, and are left with a single equation connecting T, p, v and T_c, p_c, v_c. Considerations of physical dimensions shew that it must be possible to put this equation in the form

$$\mathfrak{p} = f(\mathfrak{t}, \mathfrak{v}), \qquad \ldots\ldots(110)$$

in which the coefficients in f are independent of the nature of the gas. This is the result which has been already established for the special equations of Van der Waals and Dieterici.

Assuming that the gas-equation can be expressed in the form (110), two gases which have the same values of $\mathfrak{t}$, $\mathfrak{p}$ and $\mathfrak{v}$ are said to be in "corresponding" states. Clearly for two gases to be in corresponding states, it is sufficient for any two of the three quantities t, p and v to be the same for both.

74. It is sometimes asserted as a natural law, that when two of the quantities t, p and v are the same for two gases, then the third quantity will necessarily also be the same, and this supposed law is called the "Law of Corresponding States". This law will, however, only be true if the reduced equation of state can be put in the form of equation (110), and this in turn demands that the nature of the gas shall be specified by only two physical constants, as for instance the a and b of Van der Waals. There is of course no question that the law is true as a first approximation, because the equations of both Van der Waals and Dieterici are true as a first approximation, but if the law were generally true, all the entries in the table on p. 94 would be equal, which is obviously not the case.

The law of corresponding states can be expressed in the form that by contraction or expansion of the scales on which p and v are measured, the isothermals of all gases can be made exactly the same. This same statement can be put in the alternative form that graphs in which $\log p$, $\log \rho$ and $\log T$ are plotted against one another will be the same for all gases.

OTHER EQUATIONS OF STATE

The Empirical Equation of State of Kamerlingh Onnes

75. Following Kamerlingh Onnes, let us introduce a quantity K defined for any gas by

$$K = \frac{NRT_c}{p_c v_c},$$

a consideration of physical dimensions shewing that K must be a pure number, and let us further put

$$v_K = \frac{v}{K}.$$

With this notation, Van der Waals' equation (108) reduces to

$$\left(p + \frac{27}{64 v_K^2}\right)\left(v_K - \frac{1}{8}\right) = t, \qquad \ldots\ldots(111)$$

which can also be written in the form

$$\mathfrak{p}\mathfrak{v}_K = \frac{t}{1 - \dfrac{1}{8\mathfrak{v}_K}} - \frac{27}{64\mathfrak{v}_K}$$

$$= t\left\{1 + \frac{1}{\mathfrak{v}_K}\left(\frac{1}{8} - \frac{27}{64}\frac{1}{t}\right) + \frac{1}{64\mathfrak{v}_K^2} + \frac{1}{512\mathfrak{v}_K^3}\right.$$

$$\left. + \frac{1}{4096\mathfrak{v}_K^4} + \frac{1}{40768\mathfrak{v}_K^5} + \ldots\right\}. \quad \ldots\ldots(112)$$

As innumerable observers have found that an equation of this type is not adequate to represent the various states of a gas, Kamerlingh Onnes proposed the more general empirical form

$$\mathfrak{p}\mathfrak{v}_K = t\left\{1 + \frac{\mathfrak{B}}{\mathfrak{v}_K} + \frac{\mathfrak{C}}{\mathfrak{v}_K^2} + \frac{\mathfrak{D}}{\mathfrak{v}_K^4} + \frac{\mathfrak{E}}{\mathfrak{v}_K^6} + \frac{\mathfrak{F}}{\mathfrak{v}_K^8}\right\}, \quad \ldots\ldots(113)$$

where $\mathfrak{B}$, $\mathfrak{C}$, $\mathfrak{D}$, ... are themselves series of the form

$$\mathfrak{B} = \mathfrak{b}_1 + \frac{\mathfrak{b}_2}{t} + \frac{\mathfrak{b}_3}{t^2} + \frac{\mathfrak{b}_4}{t^4} + \frac{\mathfrak{b}_5}{t^6},$$

an expansion which contains no fewer than twenty-five adjustable coefficients. *84931*

If the law of corresponding states were true, the coefficients in the expansions (111), (112) ought to be the same for all gases.

Actually Kamerlingh Onnes[*] found that an equation of state of the form (113) can be obtained which expresses with fair accuracy the observations of Amagat on hydrogen, oxygen, nitrogen and $C_4H_{10}O$, also of Ramsay and Young on $C_4H_{10}O$ and of Young on isopentane (C_5H_{12}). The coefficients in this equation are found to be those given in the following table:

	1	2	3	4	5
$10^3\,\mathfrak{b}$	117·796	−228·038	−172·891	−72·765	−3·172
$10^4\,\mathfrak{c}$	135·580	−135·788	295·908	160·949	51·109
$10^5\,\mathfrak{d}$	66·023	−19·968	−137·157	55·851	−27·122
$10^7\,\mathfrak{e}$	−179·991	648·583	−490·683	97·940	4·582
$10^9\,\mathfrak{f}$	142·348	−547·249	508·536	−127·736	12·210

* *Encyc. d. Math. Wissenschaften*, 5 (1912), 10, p. 729, or *Comm. Phys. Lab. Leiden*, **11**, Supp. 23 (1912), p. 115.

The circumstance that the same equation of state is valid for all these six gases shews of course the wide applicability of the law of corresponding states. On the other hand, the coefficients of the equation of state of argon and helium are found to differ very substantially from those in the above table. Indeed Kamerlingh Onnes found that the isothermals of helium[*] can be represented very fairly between 100° C. and −217° C. by the equation

$$\mathfrak{p}\mathfrak{v} = NRT + \frac{NRTb - a}{\mathfrak{v}} + \frac{5}{8}\frac{NRTb^2}{\mathfrak{v}^2},$$

which is simply Van der Waals' equation adjusted by the inclusion of Boltzmann's second-order correction (§ 44). It must however be noticed that the value of T_c for helium is very low, about 5·3° abs., so that the range of temperatures studied by Prof. Kamerlingh Onnes is about from $t = 10$ to $t = 70$, and for these high values of t it is inevitable that Van der Waals' equation should in any case give a good approximation.

The Equation of State of Clausius

76. Various attempts have been made to improve Van der Waals' equation by the introduction of a few more adjustable constants, which can be so chosen as to make the equation agree more closely with experiment.

One of the best known of these improved equations is that of Clausius, namely

$$\left(p + \frac{a'}{T(v+c)^2}\right)(v - b) = NRT. \qquad \text{......(114)}$$

On putting $c = 0$, the equation becomes similar to that of Van der Waals, except that the a of Van der Waals' equation is replaced by a'/T; in other words, instead of a being constant, it is supposed to vary inversely as the temperature. For some gases such an equation is found to fit the observations better than the equation of Van der Waals.

If, however, c is not put equal to zero, but is treated as an adjustable constant and selected to fit the observations, there is found to be no tendency for c to vanish. The following table shews

[*] *Comm. Phys. Lab. Leiden,* **102** a (1907).

the values of a', b and c which are found by Sarrau* to give the best approximation to the observations of Amagat:

Gas	a'	b	c	c/b
Nitrogen	0·4464	0·001359	0·000263	0·19
Oxygen	0·5475	0·000890	0·000686	0·75
Ethylene	2·688	0·000967	0·001919	1·98
Carbon-dioxide	2·092	0·000866	0·000949	1·10

77. We have already seen that, if Van der Waals' equation were true, the law of corresponding states would follow as a necessary consequence. The reason for this is that in Van der Waals' law there are only two constants, a and b, which respectively provide the scales on which the pressure and volume can be measured in reduced coordinates.

The equation of Clausius, on the other hand, contains three separate constants, a', b and c, and of these two, namely b and c, provide different scales on which the volume can be measured: these two scales only become identical if b and c stand in a constant ratio to one another. Thus if the law of corresponding states were true, the ratio c/b would be the same for all gases. The last column of the above table shews, however, that there is not even an approximation to constancy in the values of c/b.

78. Clausius originally devised formula (114) in an effort to find a formula which should fit the observations of Andrews on carbon-dioxide. It was found that this formula, although fairly successful in the case of carbon-dioxide, was not equally successful with other gases, and Clausius then suggested the more general form

$$\left\{ p + \left(\frac{a''}{T^{n-1}} - a'''T \right) \frac{1}{(v+c)^2} \right\} (v-b) = NRT,$$

which contains five adjustable constants. For carbon-dioxide, it is found that $n = 2$ and $a''' = 0$, so that the equation reduces to (114), but for other gases n and a''' do not approximate to these values. For instance Clausius finds that for ether $n = 1·192$, for water-vapour $n = 1·24$, while to agree with observations on alcohol, n itself must be regarded as a function of the temperature

* *Comptes Rendus*, **114** (1882), pp. 639, 718, 845.

and pressure, having values which vary from 1·087 at 0° C. to 0·184 at 240° C.

Beatty and Bridgeman have suggested an equation of state, also containing five adjustable constants, of the form

$$\left[p + \frac{A}{v^2}\left(1 - \frac{a}{v}\right) \right] v = NRT\left(1 - \frac{c}{vT^3}\right)\left(1 + \frac{B}{V} - \frac{bB}{V^2}\right).$$

They find that this represents at least fourteen gases, to within an accuracy of one-half of one per cent, down to the critical point.

It is, however, obvious that there can be no finality in any of these formulae; it is possible to go on extending them indefinitely without arriving at a fully satisfactory formula, as might indeed be anticipated from the circumstance that they are purely empirical, and not founded on any satisfactory theoretical basis.

Chapter IV

COLLISIONS AND MAXWELL'S LAW

79. In § 8 we traced out the path of a single molecule under the simplest conceivable conditions. It is only in very exceptional cases that such a procedure is possible. Usually the kinetic theory has to rely on statistical calculations, based on the supposition that the number of molecules is very great indeed; the molecules are not discussed as individuals, but in crowds. In doing this, the theory encounters all the well-known difficulties of statistical methods. Indeed these confront us at the very outset, when we attempt to frame precise definitions of the simplest quantities, such as the density, pressure, etc. of a gas.

The Definition of Density

80. Let us fix our attention on any small space inside the gas, of volume Ω. Molecules will be continually entering and leaving this small space by crossing its boundary. Each time this happens, the number of molecules inside the space increases or decreases by unity. Thus the number of molecules inside the space will be represented by a succession of consecutive numbers, such as

$$n, n+1, n+2, n+1, n, n+1, n, n-1, n, n-1, n-2,$$
$$n-1, n, n+1, n, n-1, \ldots.$$

This succession of numbers will fluctuate around their average value, say n. Since the fluctuations arise from the passage of molecules over the boundary, they will be proportional to the area of this boundary, and so to the square of the linear dimensions of the space, but the number n will be proportional to the volume of the space, and so to the cube of its linear dimensions. As the volume is made larger, the fluctuations become more and more insignificant in proportion to n.

In an actual gas the volume Ω can usually be chosen of such size that it shall contain a very large number of molecules, and yet be very small compared with the scale on which the physical properties of the gas vary; a cubic millimetre is often a suitable volume. The number of molecules contained in an element of this kind, divided by the volume Ω of the element, will give the molecular density of the gas, which is usually denoted by ν. Actually ν is a fluctuating quantity, but if the volume Ω can be chosen sufficiently large, the fluctuations are inappreciable. In such a case ν may be treated as a steady quantity, and we may think of it as the number of molecules per unit volume in a specified region of the gas.

If each molecule is of mass m, and

$$\rho = m\nu,$$

ρ will measure the mass of all the molecules per unit volume; it is the "density" of the gas.

The Law of Distribution of Velocities

81. We have already seen that the molecules of a gas do not all move at the same speed; the components of velocity can have all values. For instance, when the gas has reached a steady state, they are distributed according to Maxwell's law (§ 23).

To discuss the velocities of a great number N of molecules, let us take any point as origin, and draw from it a number of separate lines, each line representing the velocity of one molecule, both in magnitude and direction. The point at the extremity of any such line will have as its coordinates u, v, w, the components of velocity of the corresponding molecule. The distribution of points obtained in this way is of course the same as the distribution of the molecules would be in space if they all started together at a point and each moved for unit time with its actual velocity. The number of points of which the coordinates lie between the limits u and $u + du$, v and $v + dv$, w and $w + dw$ may be denoted by $\tau \, du \, dv \, dw$, where τ corresponds to the ν above, being the "density" of points at the point u, v, w. The number of molecules of which the velocities lie between u and $u + du$, v and $v + dv$, w and $w + dw$ will now be $\tau \, du \, dv \, dw$. If N is the total number of molecules under consideration, it is often convenient to replace τ by Nf. If it is further necessary to specify the point u, v, w at which f is measured, we may write $f(u, v, w)$ instead of f. Because of the occurrences of collisions, $f(u, v, w)$ is a fluctuating function of u, v, w, but under suitable conditions the fluctuations will be inappreciable.

To avoid the continual repetition of these limits, let us describe a molecule of which the components of velocity lie between u and $u + du$, v and $v + dv$, w and $w + dw$ as a molecule of class A. Then the number of molecules of class A is

$$Nf(u, v, w) \, du \, dv \, dw.$$

Since the total number of molecules under consideration is N, the probability that any single molecule selected at random shall have velocity-components lying between u and $u + du$, v and $v + dv$, w and $w + dw$ will be $f(u, v, w) \, du \, dv \, dw$.

The Assumption of Molecular Chaos

82. Let us fix our attention on any small volume of the gas which contains N molecules. We may imagine that we measure the velocity of each molecule, represent it by a point u, v, w in our diagram, and so calculate $f(u, v, w)$ for all values of u, v and w. Having done this, we may proceed to some other group of N molecules and again calculate $f(u, v, w)$ for all values of u, v and w. The two sets of values of f may prove to be the same, in which case we say that the two groups of molecules have the same distribution of velocities; or obviously they may be different, as for instance if the two groups of molecules were taken from regions of the gas which were at different temperatures.

Let us now suppose that the second group of molecules is chosen from the same region of the gas as the first, but that it is chosen in a very special way. Let it consist of all the molecules which lie within a certain small distance ϵ of molecules which are themselves moving at exceptionally rapid speeds. The question arises as to whether this special way of choosing our group of molecules has any influence on the distribution of velocities. It might at first be thought, with some plausibility, that there would be some such influence—a molecule near to a second and rapidly moving molecule is more likely than others to have collided with this second very energetic molecule quite recently, in which case it might have acquired an undue amount of energy from it.

Boltzmann introduced into the kinetic theory the assumption that no such influence exists; he called this the "assumption of molecular chaos", but was unable to give any proof, or even any discussion, of its truth. We also shall make this assumption, ultimately justifying our procedure by a proof (given in Appendix IV), that the assumption is true. We shall make the assumption in the form that the chance that a molecule of class A shall be found within any small element of volume $dx\,dy\,dz$ is

$$\nu f(u, v, w)\,du\,dv\,dw\,dx\,dy\,dz, \qquad \ldots\ldots(115)$$

where f is the law of distribution of velocities for the group of molecules in a small volume Ω which includes the smaller element $dx\,dy\,dz$. We assume this to be true for every small element of volume $dx\,dy\,dz$, no matter how this is chosen.

The Changes produced by Collisions when the Molecules are Elastic Spheres

83. From the statistical point of view, the state of a gas is fully known when its density and the law of distribution of velocities are known at every point. As every collision between molecules changes the velocities of the two molecules concerned, the repeated occurrence of collisions must produce continual changes in the law of distribution of velocities. We shall first discuss these changes in a general way; ultimately we shall find that every gas may, after a sufficient time, reach a steady state, i.e. a state which is not altered by collisions, so that the density and law of distribution of velocities remain statistically the same at every point of the gas throughout all time—just as a population can attain a steady state, so that births and deaths and the normal process of growing old do not alter its statistical distribution by ages. This steady state is specified by Maxwell's law, which we have already obtained in § 23.

To begin by discussing the problem in its simplest form, let us suppose that the molecules of the gas are smooth rigid spheres, and that the physical conditions are the same at every point of the gas. Let us first imagine the gas to be in a state in which the molecular density and the law of distribution of velocities $f(u, v, w)$ are the same at every point of space. Since there is

nothing to differentiate the various regions in space, this uniformity of distribution must persist throughout all time, although
the actual form of the function f may change with the time.

84. Let us first consider the simplest possible kind of collision,
a direct head-on collision between two molecules which are
moving with velocities u and u' along a line which we may take to
be the axis of x. As in § 10, the total momentum along the axis of
x, namely $m(u + u')$, will remain the same after collision, while the
relative velocity $u' - u$ changes its sign, because of the perfect
elasticity of the molecules. It follows that the molecules merely
exchange their velocities along the axis of x; if the velocities
before collision are u, u', the velocities after collision are u', u
respectively.

Any additional velocities the molecules may have in the directions of Oy and Oz will persist unchanged after the collision, since

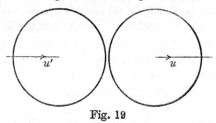

Fig. 19

there is no force to change them. Thus if the molecules had
velocities before collision

$$u, v, w \text{ and } u', v', w',$$

their velocities after collision will be

$$u', v, w \text{ and } u, v', w'.$$

By considering a number of collisions with velocities very near,
but not exactly equal to, the foregoing, we see that if the velocities
before collision lie within the limits

$$u \text{ and } u + du, \quad v \text{ and } v + dv, \quad w \text{ and } w + dw$$
$$\text{and } u' \text{ and } u' + du', \quad v' \text{ and } v' + dv', \quad w' \text{ and } w' + dw',$$

then the velocities after collision will lie within the limits

$$u' \text{ and } u' + du', \quad v \text{ and } v + dv, \quad w \text{ and } w + dw$$
$$\text{and } u \text{ and } u + du, \quad v' \text{ and } v' + dv', \quad w' \text{ and } w' + dw'.$$

In discussing such a group of collisions, we may speak of the product of the six differentials $du\,dv\,dw\,du'\,dv'\,dw'$ as the spread of the velocities. We now see that the total spread of the six components before collision, namely $du\,dv\,dw\,du'\,dv'\,dw'$, is exactly equal to the total spread after collision, namely $du'\,dv\,dw\,du\,dv'\,dw'$—a result we shall need later.

So far the collision has been supposed to take place with its line of impact along the axis of x, the relative velocity being $u'-u$. In such a case we obtain the velocities after the collision by super-posing the relative velocity along the line of impact (taken with its appropriate sign) on to the velocities before impact. Now as this rule makes no mention of the special direction (namely Ox) that we have chosen for the line of impact, it must be of universal application. So also must the other result we have obtained—that the total spread $du\,dv\,dw\,du'\,dv'\,dw'$ of the velocities b̶e̶f̶o̶r̶e̶ collision is equal to the total spread after collision.

85. Let us now remove the restriction that the line of impact is to be in the direction Ox, and consider a more general type of collision, which we shall describe as a collision of type α, and shall define by the three following conditions:

(i) One of the two colliding molecules is to be a molecule of class A, defined by the condition that the velocity-components lie within the limits u and $u+du$, v and $v+dv$, w and $w+dw$.

(ii) The second colliding molecule is to be of class B, defined by the condition that the velocity-components lie within the limits u' and $u'+du'$, v' and $v'+dv'$, w' and $w'+dw'$.

(iii) The direction of the line of impact is to have direction-cosines l, m, n, and is to lie within a specified small solid angle $d\omega$.

If the molecules all have the same diameter σ, a collision will occur each time that the centres of any two molecules come within a distance σ of one another. Let us now imagine a sphere of *radius* σ drawn round the centre of each molecule of class A. We can mark out on the surface of this sphere the solid angle $d\omega$ specified in condition (iii), thus obtaining an area $\sigma^2\,d\omega$, in a direction from the centre of the sphere of direction-cosines l, m, n. Clearly a collision of type α will occur every time that the centre of a molecule of class B comes upon this area $\sigma^2\,d\omega$.

Let the relative velocity of the two molecules at such a collision be V. In any small interval of time dt before such a collision occurs, the second molecule will move relative to the first, through a distance $V\,dt$ in the direction of V. Hence at a time dt before the collision, the centre of the second molecule must have lain on an area $\sigma^2\,d\omega$ which can be obtained by moving our original area $\sigma^2\,d\omega$ through a distance $V\,dt$ from its old position in the direction opposite to that of V. And if a collision is to take place some time *within* the interval dt, the centre of the second molecule must lie inside the cylinder which this element $\sigma^2\,d\omega$ traces out as it makes the motion just mentioned (see fig. 20).

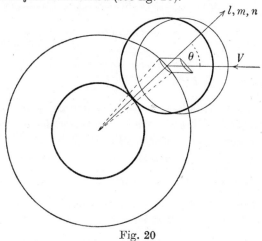

Fig. 20

The base of this cylinder is $\sigma^2\,d\omega$; its height is $V\,dt \times \cos\theta$, where θ is the angle between the direction l, m, n and that of V reversed. Hence its volume, which is the product of base and height, is $V\sigma^2\cos\theta\,d\omega\,dt$. The assumption of molecular chaos now comes into play, and we suppose that the probability that a molecule of class B shall lie within this small cylinder at the beginning of the interval dt is

$$\nu f(u', v', w')\,du'\,dv'\,dw'\,V\sigma^2\cos\theta\,d\omega\,dt. \qquad \ldots\ldots(116)$$

This, then, must also be the probability that our single molecule of class A shall experience a collision of type α within the interval of time dt.

Each unit volume of the gas contains $\nu f(u, v, w)\, du\, dv\, dw$ molecules of class A, and on multiplying this by expression (116) we obtain the total number of collisions of type α which occur in a unit volume of the gas in a time-interval dt. The number is found to be

$$\nu^2 f(u, v, w) f(u', v', w')\, V\sigma^2 \cos\theta\, du\, dv\, dw\, du'\, dv'\, dw'\, d\omega\, dt. \quad \ldots\ldots(117)$$

Each of these collisions results in one molecule leaving class A.

86. Let us next consider a type of collision which results in a molecule entering class A. We shall describe this as a collision of type β, and shall define it by the three conditions that

(i) After the collision, one of the molecules is to be of class A.

(ii) After the collision, the other molecule is to be of class B.

(iii) The direction of the line of centres at impact is to satisfy the same condition as for a collision of type α (p. 108).

In brief, the two molecules have velocities before collision which are identical with those which the two molecules of a type α collision acquire after collision, and vice versa.

Let the velocities before collision be

$\bar{u}, \bar{v}, \bar{w}$ and $\bar{u}', \bar{v}', \bar{w}'$ with spreads $d\bar{u}\, d\bar{v}\, d\bar{w}$ and $d\bar{u}'\, d\bar{v}'\, d\bar{w}'$.

Making the necessary alterations in formula (117), we see that the number of collisions of type β which occurs per unit volume in time dt is

$$\nu^2 f(\bar{u}, \bar{v}, \bar{w}) f(\bar{u}', \bar{v}', \bar{w}')\, V\sigma^2 \cos\theta\, d\bar{u}\, d\bar{v}\, d\bar{w}\, d\bar{u}'\, d\bar{v}'\, d\bar{w}'\, d\omega\, dt. \quad \ldots\ldots(118)$$

The quantities ν, σ^2, $\cos\theta$, $d\omega$ and dt stand unaltered, since these have the same value for both types of collision. Furthermore, the result obtained in § 84 shews that the total spread of velocities after collision is equal to the total spread before, so that

$$d\bar{u}\, d\bar{v}\, d\bar{w}\, d\bar{u}'\, d\bar{v}'\, d\bar{w}' = du\, dv\, dw\, du'\, dv'\, dw'$$

and formula (118) may be written in the form

$$\nu^2 f(\bar{u}, \bar{v}, \bar{w}) f(\bar{u}', \bar{v}', \bar{w}')\, V\sigma^2 \cos\theta\, du\, dv\, dw\, du'\, dv'\, dw'\, d\omega\, dt. \quad \ldots\ldots(119)$$

The Condition for a Steady State

87. We may notice that expression (119), which specifies the number of collisions of type β, is identical with expression (117), which specifies the number of collisions of type α, except that $f(\overline{u}, \overline{v}, \overline{w})f(\overline{u}', \overline{v}', \overline{w}')$ replaces $f(u, v, w)f(u', v', w')$.

If, then,

$$f(\overline{u}, \overline{v}, \overline{w})f(\overline{u}', \overline{v}', \overline{w}') > f(u, v, w)f(u', v', w'),$$

collisions of type β will be more frequent than collisions of type α, and class A of molecules will gain by collisions of these two types. If

$$f(\overline{u}, \overline{v}, \overline{w})f(\overline{u}', \overline{v}', \overline{w}') = f(u, v, w)f(u', v', w'), \quad ......(120)$$

then collisions of types α and β occur with equal frequency, so that class A neither increases nor decreases its numbers as the result of collisions of these two types. If equation (120) is true for all values of the velocity-components, then class A neither gains nor loses as the result of collisions of any type whatsoever. The same is true of every other class of molecules, so that if equation (120) is true for all velocities, each class retains the same number of molecules throughout all collisions—the gas is in a steady state.

It is, then, very important to examine whether equation (120) can be satisfied for all velocities. Since a collision changes u, v, w, u', v', w' into $\overline{u}, \overline{v}, \overline{w}, \overline{u}', \overline{v}', \overline{w}'$, the equation merely expresses that $f(u, v, w)f(u', v', w')$ remains unchanged by a collision, or again, taking logarithms, that

$$\log f(u, v, w) + \log f(u', v', w') \qquad ......(121)$$

remains unchanged at a collision.

This expression is the sum of two contributions, one from each molecule, and we immediately think of a number of quantities of which the sum is conserved at a collision. To begin with there are the energy $\frac{1}{2}m(u^2 + v^2 + w^2)$, and the three components of angular momentum (mu, mv, mw); these give four possible forms for $\log f(u, v, w)$. A fifth is obtained by taking $\log f(u, v, w)$ equal to a constant (conservation of number of molecules, i.e. mass), and it is obvious that there can be no others. For if any additional form were possible for $\log f$, there would be five equations giving $\overline{u}, \overline{v}, \overline{w}, \overline{u}', \overline{v}', \overline{w}'$ in terms of u, v, w, u', v', w', so that $\overline{u}, \overline{v}, \overline{w}, \overline{u}', \overline{v}', \overline{w}'$

would be determined except for one unknown. There must, however, be two unknowns, as the direction of the line of centres is unknown.

The general solution of equation (120) is therefore seen to be

$$\log f(u, v, w) = \alpha_1 m(u^2 + v^2 + w^2) + \alpha_2 mu + \alpha_3 mv + \alpha_4 mw + \alpha_5,$$
$$\dots\dots(122)$$

where $\alpha_1, \alpha_2, \alpha_3, \alpha_4, \alpha_5$ are constants. These may be replaced by new constants and the solution written in the form

$$\log f(u, v, w) = \alpha_1 m[(u - u_0)^2 + (v - v_0)^2 + (w - w_0)^2] + \alpha_6$$

or, if we still further change the constants,

$$f(u, v, w) = A e^{-hm[(u - u_0)^2 + (v - v_0)^2 + (w - w_0)^2]}, \qquad \dots\dots(123)$$

in which A, h, u_0, v_0, w_0 are new arbitrary constants.

If $f(u, v, w)$ is of this form, the gas will be in a steady state, i.e. the distribution of velocities will not be changed by collisions. This value of f is a slight generalisation of that previously obtained in § 23 (formula (41)).

88. Let us now examine what these various constants mean in physical terms. As the index of the exponential in formula (123) is necessarily negative or zero for all values of u, v and w, we see that $f(u, v, w)$ has its maximum value when the index is zero, i.e. when

$$u = u_0, \quad v = v_0, \quad w = w_0.$$

This shews that the velocities of most frequent occurrence are those in the immediate neighbourhood of u_0, v_0, w_0.

Further, since $(u - u_0)$ occurs only through its square, positive values of $u - u_0$ occur exactly as frequently as the corresponding negative values, so that the average value of $u - u_0$ is zero. Thus the average value of u is u_0, so that u_0, v_0, w_0 must be the components of velocity of the centre of gravity of the whole gas, while the quantities $u - u_0$, $v - v_0$, $w - w_0$, which figure in the exponential in equation (123), are simply the components of the velocity of a molecule relative to the centre of gravity of the gas.

Let us next write

$$u - u_0 = \mathsf{U},$$
$$v - v_0 = \mathsf{V},$$
$$w - w_0 = \mathsf{W},$$

so that $u = u_0 + U$, etc. We now see that the velocity of any molecule may be regarded as made up of

(i) a velocity u_0, v_0, w_0 which is the same for all molecules, being the velocity of the centre of gravity of the gas, and

(ii) a velocity U, V, W relative to the centre of gravity of the gas, which of course is different for each molecule.

We may speak of the former velocity as the "mass-velocity" of the gas, and of the latter velocity as the "thermal" velocity of the molecule, since it arises from the thermal agitation of the gas. From formula (123), the law of distribution of thermal velocities is

$$f(U, V, W) = Ae^{-hm(U^2+V^2+W^2)}. \qquad \ldots\ldots(124)$$

This brings us back to Maxwell's law, which we obtained in § 23. We see that h is identical with the h of § 23, which was there found to have the value $1/2RT$.

The right-hand member of this equation shews that the probability that the thermal velocity of a molecule shall have components which lie within the limits U and $U + dU$, V and $V + dV$, W and $W + dW$ is

$$Ae^{-hm(U^2+V^2+W^2)} \, dU \, dV \, dW.$$

If we integrate this from

$$U = -\infty \text{ to } +\infty, \quad V = -\infty \text{ to } +\infty \quad \text{and} \quad W = -\infty \text{ to } +\infty,$$

we obtain the total probability that the molecule shall have velocity-components lying somewhere within these limits. But as every molecule must have thermal velocities which lie somewhere within these limits, this total probability must be equal to unity, so that we must have

$$A \int\int\int_{-\infty}^{+\infty} e^{-hm(U^2+V^2+W^2)} \, dU \, dV \, dW = 1.$$

The value of the integral is easily found to be $(\pi/hm)^{\frac{3}{2}}$ (Appendix VI, p. 306), so that

$$A = \left(\frac{hm}{\pi}\right)^{\frac{3}{2}}.$$

Any one of the N molecules has an amount

$$\tfrac{1}{2}m(\mathsf{U}^2 + \mathsf{V}^2 + \mathsf{W}^2)$$

of thermal energy, so that the total thermal energy of the gas is obtained by summing the quantity over all the N molecules.

The number of molecules having a thermal velocity of which the components lie between the limits

$$\mathsf{U} \text{ and } \mathsf{U} + d\mathsf{U}, \quad \mathsf{V} \text{ and } \mathsf{V} + d\mathsf{V}, \quad \mathsf{W} \text{ and } \mathsf{W} + d\mathsf{W}$$

is

$$NAe^{-hm(\mathsf{U}^2 + \mathsf{V}^2 + \mathsf{W}^2)} d\mathsf{U}\, d\mathsf{V}\, d\mathsf{W},$$

and the contribution which these molecules make to the total thermal energy is

$$\tfrac{1}{2}mNA(\mathsf{U}^2 + \mathsf{V}^2 + \mathsf{W}^2)\, e^{-hm(\mathsf{U}^2 + \mathsf{V}^2 + \mathsf{W}^2)} d\mathsf{U}\, d\mathsf{V}\, d\mathsf{W}.$$

Summing this over all values of $\mathsf{U}, \mathsf{V}, \mathsf{W}$, we find the total thermal energy of the N molecules to be

$$\tfrac{1}{2}mNA \int\limits_{-\infty}^{+\infty}\!\!\!\int\!\!\!\int (\mathsf{U}^2 + \mathsf{V}^2 + \mathsf{W}^2)\, e^{-hm(\mathsf{U}^2 + \mathsf{V}^2 + \mathsf{W}^2)} d\mathsf{U}\, d\mathsf{V}\, d\mathsf{W}.$$

The value of the integral is easily found to be

$$\frac{3}{2}\sqrt{\frac{\pi^3}{h^5 m^5}},$$

so that the total thermal energy is

$$\tfrac{3}{4}mNA\sqrt{\frac{\pi^3}{h^5 m^5}} = \frac{3N}{4h}.$$

The average thermal energy of a molecule is accordingly $3/4h$, or $\tfrac{3}{2}RT$, in agreement with formula (17).

Steady State in a Mixture of Gases

89. A slightly more complex problem arises when the gas consists of a mixture of molecules of two different kinds. Let us suppose that the molecules of both kinds are hard smooth spheres; let $m_1, \sigma_1, \nu_1, f_1$ refer to the first kind of molecule, while $m_2, \sigma_2, \nu_2, f_2$ refer to the second kind. We proceed as before and obtain, in place of formula (117),

$$\nu_1 \nu_2 f_1(u, v, w) f_2(u', v', w')\, V\, [\tfrac{1}{2}(\sigma_1 + \sigma_2)]^2$$
$$\times \cos\theta\, du\, dv\, dw\, du'\, dv'\, dw'\, d\omega\, dt$$

for the number of collisions of type α in which the second molecule is supposed to be of the second kind.

If the gas is to be in a steady state, there must be no net loss or gain to molecules of class A through collisions with molecules of the second kind. The condition for this is found to be that (in place of equation (120))

$$f_1(\bar{u}, \bar{v}, \bar{w}) f_2(\bar{u}', \bar{v}'\, \bar{w}') = f_1(u, v, w) f_2(u', v', w'), \qquad \ldots\ldots (125)$$

and we see, just as before, that $\log f_1$ and $\log f_2$ must represent quantities which are conserved at collision. Again we find that $\log f_1$ and $\log f_2$ may be molecular energies, or components of momentum, or constants, or a combination of all five, so that finally f_1 and f_2 must be of the form

$$f_1(u, v, w) = A_1 e^{-hm_1[(u-u_0)^2 + (v-v_0)^2 + (w-w_0)^2]},$$
$$f_2(u, v, w) = A_2 e^{-hm_2[(u-u_0)^2 + (v-v_0)^2 + (w-w_0)^2]},$$

where h, u_0, v_0, w_0 are the same in both formulae.

The fact that h is the same for the two kinds of gases shews that in the steady state the average kinetic energy of the two kinds of molecules is the same. This is a special case of the general theorem of equipartition of energy to which we have already referred (§ 11).

Remembering that the absolute temperature in a gas is measured by the average kinetic energy of its molecules, we see that when two gases are mixed they will in time reach a steady state in which the temperature is the same for both.

THE MAXWELLIAN DISTRIBUTION

90. We may notice first that the components U, V, W of thermal velocity enter separately into Maxwell's law. To be more precise, if we write

$$\phi(U) = A^{\frac{1}{3}} e^{-hmU^2} = \left(\frac{hm}{\pi}\right)^{\frac{1}{3}} e^{-hmU^2}, \qquad \ldots\ldots (126)$$

then equation (124), which expresses Maxwell's law, can be written in the form

$$f(U, V, W)\, dU\, dV\, dW = [\phi(U)\, dU]\, [\phi(V)\, dV]\, [\phi(W)\, dW].$$
$$\ldots\ldots (127)$$

This shews that the probability of a molecule having a velocity of components U, V, W is the product of three independent probabilities, one of which depends only on U, one only on V and the third only on W. It follows that the probability of a molecule having any specified velocity in the direction of Ox is independent of the velocity-components of the molecule in other directions, being in fact determined by the formula

$$\phi(U)\, dU = \left(\frac{hm}{\pi}\right)^{\frac{1}{2}} e^{-hmU^2}\, dU.$$

To recapitulate, we have now found that equation (124) represents a steady state in which the law of distribution of thermal velocities (U, V, W) is given by the formula

$$f(U, V, W) = \left(\frac{hm}{\pi}\right)^{\frac{3}{2}} e^{-hm(U^2+V^2+W^2)}, \qquad \ldots\ldots(128)$$

while the law of distribution for a single component is

$$\phi(U) = \left(\frac{hm}{\pi}\right)^{\frac{1}{2}} e^{-hmU^2}. \qquad \ldots\ldots(129)$$

Equation (128), which expresses Maxwell's law of distribution, was first given in his 1859 paper. The proof by which he sought to establish the law is now generally agreed to have been unsound. It began by assuming, without any attempted justification, that the distributions of U, V, W were independent. This proof is reproduced in Appendix I (p. 296) on account of its historic interest. The proof given in the present chapter is a modification of one originally given by Boltzmann and Lorentz, but this also proceeded on the basis of an unjustified assumption, namely that of molecular chaos (§ 82). In respect of logical completeness, then, there is little to choose between the two proofs, but the proof just given has the merit of giving a vivid picture of the physical processes at work in a gas, besides providing a number of formulae which are otherwise useful.

The investigations of Boltzmann and Lorentz went further than this. Equation (123) contains the most general solution of equation (120), but it has not been shewn that it represents all possible steady states. Boltzmann and Lorentz proved this; their proof is given in Appendix II, and so will be assumed henceforth. Thus by giving different values to the constants h, u_0, v_0, w_0 in equation (123) we obtain all the steady states which are possible for a gas. It is further proved in Appendix II that a gas which is left to itself will gradually approach to one or other of these steady states, and must finally attain it after sufficient time has elapsed.

A more complete proof of all this, which involves neither Maxwell's assumption of the independence of U, V and W nor the assumption of molecular chaos, is given in Appendix IV (p. 301).

Thermal Motions

91. Let us next suppose that the thermal velocity U, V, W of a molecule is of amount c, and that its direction lies within a small solid angle $d\omega$. Then $U^2 + V^2 + W^2 = c^2$, and by a well-known transformation of coordinates

$$dU\, dV\, dW = c^2\, dc\, d\omega.$$

The law of distribution now assumes the form

$$Ae^{-hmc^2}c^2\, dc\, d\omega.$$

The direction of U, V, W only enters this formula through the differential $d\omega$, so that the chance of this direction lying within any solid angle is exactly proportional to the magnitude of the solid angle. In other words, all directions are equally likely, as indeed we should expect. For when the gas has no mass-motion there is no reason why thermal velocities should favour one direction more than another. When the gas has a mass-motion, the principle of relativity takes charge of the situation, and shews that the distribution of thermal velocities must be the same as when the gas has no mass-motion.

If we integrate over all possible directions for $d\omega$, the law of distribution becomes

$$4\pi Ae^{-hmc^2}c^2\, dc. \qquad\qquad \text{......(130)}$$

In fig. 21 the thin line shews the graph of the curve

$$y = e^{-x^2},$$

while the thick line is the graph of the curve

$$y = 2x^2 e^{-x^2},$$

the factor 2 being introduced in this latter curve to make the areas of the two curves the same. The thin curve gives a graphical representation of the distribution of any single component (U, V or W) of molecular velocity while the thick gives the same for the total molecular velocity c.

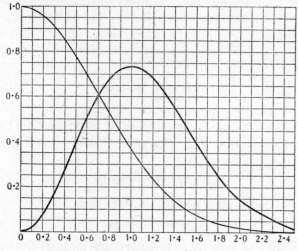

Fig. 21

On the thin curve the maximum ordinate occurs at $y = 0$, shewing that the most frequent value of U is $U = 0$. The value of y falls off rapidly as x increases, shewing that large values of U are rare. The average numerical value of U is

$$\int_0^\infty U e^{-hmU^2} dU \Big/ \int_0^\infty e^{-hmU^2} dU = \frac{1}{(\pi hm)^{\frac{1}{2}}} = \frac{0 \cdot 5642}{(hm)^{\frac{1}{2}}},$$

and so is represented by the ordinate of $x = 0 \cdot 5642$ in fig. 21. Molecules with more than four times this velocity are represented by the area of that part of the thin curve which lies to the right of the ordinate $x = 2 \cdot 257$, and this is seen to be very small indeed. Thus it is very rare for any component of molecular velocity to have as much as four times the average value.

The thick curve exhibits the distribution of c among the molecules of the gas and here the maximum ordinate is that of $x = 1$, so that the most frequent value of c is $1/(hm)^{\frac{1}{2}}$. The graph shews that values of c which are more than $2\frac{1}{2}$ times the most frequent value of c are very rare. The average value of c, which we shall denote by $\bar{c}$, is

$$\bar{c} = \int_0^\infty c^3 e^{-hmc^2} dc \bigg/ \int_0^\infty c^2 e^{-hmc^2} dc = \frac{2}{(\pi h m)^{\frac{1}{2}}} = \frac{1 \cdot 1284}{(hm)^{\frac{1}{2}}}$$

and so is just double the average value of any single coordinate U, V or W.

The mean value of c^2 may be denoted by $\overline{c^2}$, but is more commonly denoted by C^2. Its value is readily found to be

$$C^2 = \frac{3}{2hm} = \frac{1 \cdot 5}{hm},$$

so that

$$\bar{c} = 0 \cdot 921 C \quad \text{and} \quad C = 1 \cdot 086 \bar{c}.$$

With C^2 defined in this way, the average thermal energy of a molecule is $\frac{1}{2}mC^2$, and the thermal energy per unit volume is $\frac{1}{2}m\nu C^2$ or $\frac{1}{2}\rho C^2$; it is the same as though the whole gas were moving forward through space with a velocity C, or $1 \cdot 086 \bar{c}$.

Mass-Motion and Molecular-Motion

92. We have seen that in the most general "steady state" possible, the motion consists of a thermal motion compounded with a mass-motion. The mass-motion has velocity components u_0, v_0, w_0, while the thermal motion has velocity components $u - u_0, v - v_0, w - w_0$, which we have denoted (p. 112) by U, V, W. The kinetic energy per unit volume of the gas is

$$\Sigma \tfrac{1}{2} m (u^2 + v^2 + w^2),$$

and since

$$\Sigma \mathsf{U} = \Sigma \mathsf{V} = \Sigma \mathsf{W} = 0,$$

we have

$$\Sigma \tfrac{1}{2} m (u^2 + v^2 + w^2) = \tfrac{1}{2} m \Sigma \{(\mathsf{U} + u_0)^2 + (\mathsf{V} + v_0)^2 + (\mathsf{W} + w_0)^2\}$$
$$= \tfrac{1}{2} m \Sigma (\mathsf{U}^2 + \mathsf{V}^2 + \mathsf{W}^2 + u_0^2 + v_0^2 + w_0^2)$$
$$= \tfrac{1}{2} m \nu (C^2 + u_0^2 + v_0^2 + w_0^2)$$
$$= \tfrac{1}{2} \rho (u_0^2 + v_0^2 + w_0^2 + C^2),$$

where ρ is the mass-density of the gas, given by $\rho = m\nu$,

This shews that the total energy of a gas may be regarded as the sum of the energies of its mass-motion and its thermal motion.

Let us suppose that a vessel containing gas, which has so far been moving with a velocity of which the components are u_0, v_0, w_0, is suddenly brought to a standstill. This will of course destroy the steady state of the gas, but after a sufficient time the gas will assume a new and different steady state. The mass-velocity of this steady state will obviously be *nil*, and the energy wholly molecular. The individual molecules have not been acted upon by any external forces except in their impacts with the containing vessel, and these leave their energy unchanged. The new molecular energy is therefore equal to the former total energy. This enables us to determine the new steady state. In the language of the older physics, one would say that by suddenly stopping the forward motion of the gas the kinetic energy of this motion had been transformed into heat. In the language of the kinetic theory we say that the total kinetic energy has been redistributed, so as now to be wholly molecular.

An interesting region of thought, although one outside the domain of pure kinetic theory, is opened up by the consideration of the processes by which this new steady state is arrived at. To examine the simplest case, let us suppose the gas to be contained in a cubical box, and to have been moving originally in a direction perpendicular to one of the sides. The hydrodynamical theory of sound is capable of tracing the motion of the gas throughout all time, subject of course to the assumptions on which the theory is based. The solution of the problem obtained from the hydrodynamical standpoint is that the original motion of the gas is perpetuated in the form of plane waves of sound in the gas, the wave fronts all being perpendicular to the original direction of motion. This solution is obviously very different from that arrived at by the kinetic theory. For instance, the solution of hydrodynamics indicates that the original direction of motion remains differentiated from other directions in space through all time, whereas the solution of the kinetic theory indicates that a state is soon attained in which there is no differentiation between directions in space.

The explanation of the divergence of the two solutions is naturally to be looked for in the differences of the assumptions made. The conception of the perfect non-viscous fluid postulated by hydrodynamics is an abstract ideal which is logically inconsistent with the molecular constitution of matter postulated by the kinetic theory. Indeed in a later part of the book we shall find that molecular structure is inconsistent with non-viscosity; and shall be able to shew that the actual viscosity of gases is simply and fully accounted for by their molecular structure. If we introduce viscosity terms into the hydrodynamical discussion, the energy of the original motion becomes "dissipated" by viscosity. On the kinetic theory view, this energy has been converted into molecular motion. In fact the kinetic theory enables us to trace as molecular motion energy which other theories are content to regard as lost from sight.

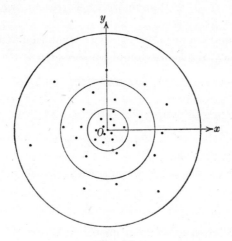

Fig. 22

93. Maxwell's law of distribution of velocities has an obvious similarity to the well-known law of trial and error. If a marksman fires shots at a target, the result will look somewhat as in fig. 22. The marksman tries, in firing each shot, to make both the x and the y coordinates as small as possible. Subject to certain assumptions, which we need not discuss here, Gauss shewed that the two efforts may be treated as independent. The chance that the x coordinate shall be between x and $x + dx$ is

$$\sqrt{\frac{\kappa}{\pi}}\, e^{-\kappa x^2} dx, \qquad \qquad \ldots \ldots (131)$$

where κ is a constant which measures the skill of the marksman. If the marksman is skilled, small values of x are frequent, so that κ is large; the more skilled the marksman, the greater the value of κ, this becoming infinite for an ideal marksman of perfect skill. Formula (131) is similar to Maxwell's formula (126) for the chance of a velocity component between U and $\mathsf{U} + d\mathsf{U}$, except that κ replaces hm, which, as we have seen, is inversely proportional to C^2 and so to the absolute temperature of the gas. Thus we may imagine Maxwell's formula arrived at by all the molecules aiming at zero for each component velocity. The total of their errors of marksmanship provides the total thermal energy of the gas, so that a high temperature corresponds to bad marksmanship and conversely.

When a marksman aims at a two-dimensional target, as in fig. 22, the law of distribution of his shots is

$$\frac{\kappa}{\pi}\, e^{-\kappa(x^2 + y^2)} dx\, dy,$$

or, transforming to plane polar coordinates r, θ, is

$$\frac{\kappa}{\pi}\, e^{-\kappa r^2} r\, dr\, d\theta.$$

If the target is three-dimensional, the corresponding law becomes
$$\left(\frac{\kappa}{\pi}\right)^{\frac{3}{2}} e^{-\kappa(x^2 + y^2 + z^2)} dx\, dy\, dz,$$

or, in three-dimensional polar coordinates r, θ, ϕ,

$$\left(\frac{\kappa}{\pi}\right)^{\frac{3}{2}} e^{-\kappa r^2} r^2\, dr \sin\theta\, d\theta\, d\phi,$$

which is now identical with Maxwell's law (§ 91). Just as (in fig. 22) the marksman aims at a radial distance $r = 0$, so here the molecules aim at a total velocity $c = 0$; the average velocity is not, however, $\bar{c} = 0$, but as we have already seen $c = \dfrac{1\cdot 1284}{(hm)^{\frac{1}{2}}}$.

94. Although Maxwell's law is comparatively simple in itself, its introduction into a physical problem often leads to complicated, and even intractable, mathematics. It is sometimes permissible to avoid the introduction of Maxwell's law, and obtain a rough approximation to the solution of a physical problem by assuming that all the molecules have exactly the same velocity, and in order that this velocity may be consistent with the actual values of the pressure and of the kinetic energy, this uniform velocity must be supposed to be C.

Some idea of the amount of error involved in this approximation may be obtained from a study of fig. 21. Since $C^2 = 3/(2hm)$, and hm is taken equal to unity in drawing the curves of fig. 21, the approximation amounts to the assumption that the whole area of the curve is collected close to the abscissa

$$x = \sqrt{\frac{3}{2}} = 1 \cdot 225.$$

The graph shews that the approximation is a very rough one.

95. It is sometimes required to know how many of the molecules of a gas have a speed greater than or less than a given speed c_0. Out of a total of N molecules the number which have a speed in excess of c_0 is, by formula (130),

$$4\pi N \left(\frac{hm}{\pi}\right)^{\frac{3}{2}} \int_{c_0}^{\infty} e^{-hmc^2} c^3 \, dc,$$

and, on integrating by parts, this becomes

$$2N \left(\frac{hm}{\pi}\right)^{\frac{1}{2}} \left\{ \int_{c_0}^{\infty} e^{-hmc^2} \, dc + c_0 e^{-hmc_0^2} \right\}.$$

In terms of the probability integral, or error function defined by

$$\operatorname{erf} x = \frac{2}{\sqrt{\pi}} \int_x^{\infty} e^{-x^2} dx,$$

this number becomes

$$N \left(\operatorname{erf} x + \frac{2}{\sqrt{\pi}} x e^{-x^2} \right), \qquad \ldots\ldots(132)$$

where

$$x = (hm)^{\frac{1}{2}} c_0 = \sqrt{\frac{3}{2}} \left(\frac{c_0}{C}\right).$$

From a table of values of erf x, it is easy to calculate the number of molecules either having velocity greater than any value c_0, or within any range c_0 to c_1. The general run of the numbers to be expected can often be sufficiently seen by an inspection of the curves of fig. 21, but for exact calculations the table given in Appendix V will be required. The values of $1 - \text{erf } x$ are given in the fourth column of this table.

Experimental Verification of Maxwell's Law

96. The more general discussion given in Chap. X below will shew that Maxwell's law is not only applicable to the molecules of gases, but also to those of liquids, to the atoms of solids, to the free electrons in solids and even to the Brownian movements of particles suspended in a liquid or a gas. Although the law is of such outstanding importance in many branches of physics, it is only in recent years that experimental confirmation of its truth has been obtained.

The first confirmation of the law was provided by the electrons escaping from a hot metallic filament. In the interior of a conducting filament, electrons are moving in all directions. Those which reach the surface with a speed above a certain limit may overcome the restraining forces at the surface and pass into outer space, much as molecules escape from the surface of a liquid by evaporation. The speed with which any individual electron leaves the surface provides a record of the speed with which it was originally travelling inside the metal. By putting various retarding potentials on the escaping electrons, it is possible to find what proportion of the electrons are travelling with speeds above a succession of assigned limits, and so to plot out the law of distribution of velocities, both outside and inside the filament.

Experiments of this general type were originally performed by O. W. Richardson and F. C. Brown,* and later by Schottky,† Sih Ling Ting‡ and J. H. Jones.§

---

* *Phil. Mag.* **16** (1908), pp. 353, 890; **18** (1909), p. 681.
† *Ann. d. Phys.* **44** (1914), p. 1011.
‡ *Proc. Roy. Soc.* **98 A** (1921), p. 374.
§ *Proc. Roy. Soc.* **102 A** (1923), p. 734.

The original experiments of Richardson and Brown were concerned only with the electrons escaping from hot platinum, and shewed that Maxwell's law was obeyed with considerable accuracy. Then Schottky obtained similar results with carbon and tungsten, except that all velocities appeared to be substantially higher than could be reconciled with the temperature of the filament. The work of Jones cleared up these difficulties, and finally Germer,* making an extensive series of experiments at temperatures which ranged from 1440° abs. to 2475° abs., shewed that the speeds of the electrons not only satisfied Maxwell's law with accuracy, but also that their absolute magnitudes agreed exactly with the temperature of the hot filament.

97. Meanwhile, attacks were being made on the far more difficult problem of testing the law directly for molecular velocities. In 1920, Stern† planned an arrangement, a development of that of Dunoyer (§5), by which he hoped to be able to measure molecular velocities directly. Fig. 23 shews a cross-section of the apparatus. At its centre is a platinum wire W heavily coated with silver. Surrounding this is a cylindrical drum D containing a narrow slit S, and still farther out another concentric drum PP'. The whole apparatus can be rotated as a rigid body about W.

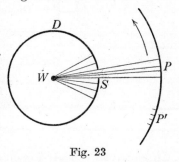

Fig. 23

The complete apparatus is put inside an enclosure in which a good vacuum can be obtained. If the wire W is now heated to a suitable temperature, atoms of silver are emitted in profusion, and if the gas-pressure is sufficiently low, so that the free path is very great compared with the dimensions of the apparatus, some of these silver molecules will pass through the slit S and form a deposit on the outer drum PP'. If the apparatus is at rest, this deposit will of course be formed at a point P exactly opposite S, but if the whole apparatus is rapidly rotating, the

* *Phys. Rev.* **15** (1925), p. 795.

† *Zeitschr. f. Phys.* **2** (1920), p. 49; **3** (1920), p. 417, and *Phys. Zeitschr.* **21** (1920), p. 582.

outer cylinder moves on through an appreciable distance while the molecules are crossing from the slit S, so that the deposit is formed farther back than if there were no rotation. Atoms which move at different speeds will of course form deposits at different places, and by measuring the position and intensities of the various deposits, it is possible to calculate the distribution of velocities in the atoms by which the deposits were formed. Unfortunately the apparatus did not prove sensitive enough to give very convincing results.

98. In 1927 E. E. Hall* designed a new, and more sensitive, form of the same apparatus, which is shewn diagrammatically in fig. 24.

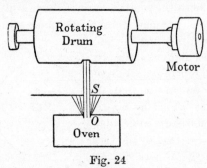

Fig. 24

An electrically heated oven represented at the bottom of the diagram contains a vapour. Molecules emerge through the opening O, moving in all directions, and the slit S selects a beam of parallel moving molecules. These move in the direction OS at different speeds which ought, as we believe, to conform to Maxwell's law. Above the slit S is a rotating drum, with a slit in its side which, once in each revolution, comes directly above the fixed slit S. If the drum were rotating very slowly, molecules would, on each of these occasions, pass through the slit and form a deposit on the point of the drum opposite to the slit. If, however, the drum rotates so fast that the speed of its surface is comparable with that of the molecules, the drum has turned through an appreciable angle before the molecule strikes its farther side, so that the point of deposition of the molecule is displaced by a distance which will be inversely proportional to the velocity of

* See Loeb, *Kinetic Theory of Gases* (2nd ed. 1934), pp. 132, 136.

the molecule. If the drum has a radius of 6 inches and rotates 100 times per second, the speed of its surface is 314 feet per second, and a molecule which is travelling at 1000 feet per second will take $\frac{1}{500}$ second to cross the drum. In this interval the surface of the drum will move

$$\tfrac{314}{500} \text{ feet or } 7\tfrac{1}{2} \text{ inches,}$$

so that the point of deposition will be displaced this far from the point opposite the slit. In this way we obtain a sort of velocity-spectrum of the speeds of the molecules of the vapour.

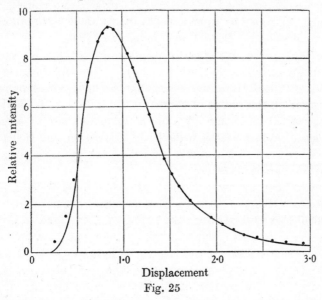

Displacement

Fig. 25

I. F. Zartman* and C. C. Ko† experimented with bismuth vapour in this way, but the velocity spectrum they obtained, shewn in fig. 25, does not entirely agree with that to be expected from Maxwell's law. Ko finds, however, that his spectrum agrees with that to be expected if the vapours consisted of a mixture of molecules of Bi, Bi_2, and Bi_3, in the proportion at 827° C. of 44 : 54 : 2, the molecules of each kind obeying Maxwell's law. In this way a rather indirect proof of Maxwell's law is obtained.

* *Phys. Rev.* **37** (1931), p. 383.
† *Journal Franklin Institute*, **217** (1934), p. 173.

99. More recently, V. W. Cohen and A. Ellett,[*] experimenting with various alkali atoms in a similar manner, have obtained a very convincing confirmation of Maxwell's law.

The same experimenters[†] have also studied the velocities in a beam of potassium atoms which had been scattered by a crystal of magnesium oxide. They found that these conformed to Maxwell's law, the absolute values being those appropriate to the temperature of the crystal, independently of the origin of the beam of atoms before striking the crystal. This suggests that the atoms were adsorbed by the crystal surface, and subsequently re-emitted from it by a process of evaporation, in accordance with the suggestion of Maxwell and Langmuir (§ 34).

In both these last experiments, the atoms were sorted out, according to their velocities of motion, by passing them through a magnetic field; the atoms, being charged, had their paths curved by the field, the radius of curvature being exactly proportional to the speed of motion. A more complicated way of magnetic sorting of atoms has been devised by Meissner and Scheffers,[‡] who have verified Maxwell's law with great precision for atoms of lithium and potassium.

100. These magnetic methods are useful only for sorting out electrically charged atoms; a magnetic field has no influence on molecules or neutral atoms. In 1926 Stern[§] proposed to sort out these by a method rather like the toothed wheel method which Fizeau used to measure the velocity of light. In 1929 Lammert[||] constructed an apparatus on these lines and obtained a very satisfactory confirmation of Maxwell's law for mercury atoms.

In fig. 26, O contains mercury vapour. A slit in this and a second slit S limit the emerging mercury atoms to a parallel moving beam. After passing through S, the atoms of the beam encounter in succession two wheels W_1, W_2, both mounted on the same shaft, and rapidly rotating. Each wheel has a slit in it, of the same width as the slit S. Once in each revolution of the wheel

[*] *Phys. Rev.* **52** (1937), p. 502.
[†] Ellett and Cohen, *Phys. Rev.* **52** (1937), p. 509.
[‡] *Phys. Zeitschr.* **34** (1933), p. 173.
[§] *Zeitschr. f. Phys.* **39** (1926), p. 751.
[||] *Zeitschr. f. Phys.* **56** (1929), p. 244.

W_1, the whole beam of molecules passes through the slit in W_1, and impinges on the wheel W_2. The slit in W_2 is not parallel to that in W_1 but is set to be at an angle of about 2 degrees behind it. Thus those molecules which travel exactly the distance between W_1 and W_2, while the wheels are turning through 2 degrees, will pass through the slit in W_2; molecules which travel at other speeds are either too late or too early to get through. By running the wheels at different speeds, it is possible to pick out the molecules which travel at any desired speed; these are allowed to form a deposit on a transparent screen, and from the density of this deposit the number of molecules can be estimated. In this way Lammert shewed that Maxwell's law was satisfied to a considerable degree of accuracy.

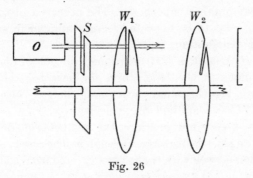

Fig. 26

Kappler obtained yet another confirmation of the law by the use of the miniature torsion-balance already described in § 5. When the arm is immersed in gas of very low density, the impacts of the separating molecules can be detected and analysed statistically. The results obtained indicate that the velocities of the individual molecules conform to Maxwell's law.

101. A certain amount of confirmation of Maxwell's law is provided by astrophysical observation. The atmosphere of a star consists of a mixture of atoms of different kinds moving with all possible velocities. If an atom of hydrogen is moving towards the earth with a speed v, and emitting the line $H\alpha$, its light when received on earth and resolved in a terrestrial spectroscope will not be found in the normal position of the $H\alpha$ line, but will be seen to be displaced through a wave-length $d\lambda$ given by

$$\frac{d\lambda}{\lambda} = \frac{v}{c},$$

in accordance with the usual Doppler-effect. As the atoms have a very wide range of speeds, the line they produce is of substantial breadth, and the distribution of intensity in this line will again provide a sort of velocity spectrum of the moving atoms. Once again, however, this gives no direct proof of Maxwell's law, since the atoms which emit the light are not all at the same depth in the star's atmosphere; the light we receive comes from places at different temperatures, so that again the velocity spectrum does not come from a single Maxwellian distribution of velocities, but from a great number superposed. The observed distribution is found to be that to be expected from a number of superposed Maxwellian distributions.

102. The objections just mentioned do not apply to a gas made luminous in the laboratory, for here the temperature, pressure, etc. are the same throughout the gas. Ornstein and van Wyck* have studied the velocity spectrum emitted by helium gas at very low pressure, the gas being made luminous by passing an electric discharge through it. The observed velocity spectrum agreed well with that to be expected from a Maxwellian distribution, but at a temperature some 50° C. above that of the gas in the tube. They thought the difference might be attributed to the general heating of the gas by the discharge.

* *Zeitschr. f. Phys.* 78 (1932), p. 734.

Chapter V

THE FREE PATH IN A GAS

103. We have already seen how viscosity and conduction of heat can be explained in terms of the collisions of gas molecules, and of the free paths which the molecules describe between collisions. As a preliminary to a more detailed study of these and other phenomena, the present chapter deals with various general problems associated with the free path, the molecules again being supposed, for simplicity, to be elastic spheres.

Number of Collisions

104. Let us begin by making a somewhat detailed study of the collisions which occur in a gas when this is in a steady state, so that the velocities are distributed according to Maxwell's law. It is clear that the occurrence of collisions cannot be affected by any mass-motion the gas may have as a whole, so that we may disregard this, put $u_0 = v_0 = w_0 = 0$, and suppose the law of distribution of velocities to be

$$f(u, v, w) = \left(\frac{hm}{\pi}\right)^{\frac{3}{2}} e^{-hmc^2}, \qquad \ldots\ldots(133)$$

where $c^2 = u^2 + v^2 + w^2$.

In §85 we calculated the frequency of collisions of various kinds in a gas in which the molecules were similar spheres, all of diameter σ. We found the number of collisions of class α occurring per unit volume per unit time to be

$$\nu^2 f(u, v, w) f(u', v', w') \, V\sigma^2 \cos\theta \, du \, dv \, dw \, du' \, dv' \, dw' \, d\omega.$$

If we replace $f(u, v, w)$ and $f(u', v', w')$ by the special values appropriate to the steady state, this number of collisions becomes

$$\nu^2 \left(\frac{hm}{\pi}\right)^3 e^{-hm(u^2+v^2+w^2+u'^2+v'^2+w'^2)} V\sigma^2 \cos\theta \, du \, dv \, dw \, du' \, dv' \, dw' \, d\omega.$$

$$\ldots\ldots(134)$$

Here V is the relative velocity at collision, given by

$$V^2 = (u-u') + (v-v') + (w-w')^2.$$

9-2

105. Expression (134) depends only on the velocities at collision, except for the factor $\sigma^2 \cos\theta \, d\omega$. Clearly $\sigma^2 \, d\omega$ is an element of area on a sphere of radius σ, say P in fig. 27, and $\sigma^2 \cos\theta \, d\omega$ is its projection Q on a plane AB at right angles to V. We may in fact replace $\sigma^2 \cos\theta \, d\omega$ by dS, an element of area in this plane; we then see that the projection is equally likely to be at all points inside a circle of radius σ, as is of course otherwise obvious.

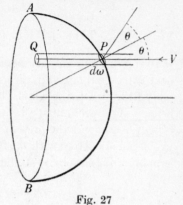

Fig. 27

In terms of the usual polar coordinates θ, ϕ, the factors which involve the angles in expression (134) may be rewritten in the form

$$\cos\theta \, d\omega = \sin\theta \cos\theta \, d\theta \, d\phi = \tfrac{1}{4}\sin 2\theta \, d(2\theta) \, d\phi. \quad \dots\dots(135)$$

When collision occurs, the relative velocity along the line of impact is reversed, while that at right angles to this line persists unchanged. Thus the relative velocity is bent through an angle 2θ, and the polar coordinates of its direction after collision are 2θ, ϕ. Formula (135) now shews that all directions are equally likely for the relative velocity after collision, a result first given by Maxwell.

106. Integrating expression (134) with respect either to dS or to θ and ϕ, we replace $\sigma^2 \cos\theta \, d\omega$ by $\pi\sigma^2$, and find that the number of collisions at all angles of incidence is

$$\pi\nu^2 \left(\frac{hm}{\pi}\right)^3 e^{-hm(u^2+v^2+w^2+u'^2+v'^2+w'^2)} V\sigma^2 \, du \, dv \, dw \, du' \, dv' \, dw'.$$

$$\dots\dots(136)$$

This is the number of collisions per unit time per unit volume of gas in which the two molecules belong to what we have described (§ 85) as class A and class B respectively—that is to say, the first molecule has a velocity of which the components lie within the limits u and $u+du$, v and $v+dv$, w and $w+dw$, while the second has a velocity of which the components lie within the limits u' and $u'+du'$, v' and $v'+dv'$, w' and $w'+dw'$.

Let us fix our attention for the moment on the x-components of velocity, namely u and u'. These enter formula (136) through the relative velocity V, and also through the factor

$$e^{-hm(u^2+u'^2)}\,du\,du'. \qquad \ldots\ldots(137)$$

We can plot values of u and u' in a plane diagram as shewn in fig. 28. We can also plot them in the same diagram with the axes turned through an angle of $45°$, the new coordinates ξ, ξ' now being given by

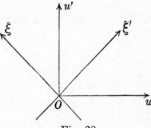

Fig. 28

$$\xi = \frac{1}{\sqrt{2}}(u'-u), \quad \xi' = \frac{1}{\sqrt{2}}(u'+u).$$

Transforming to these new coordinates, we find that

$$e^{-hm(u^2+u'^2)}\,du\,du' = e^{-hm(\xi^2+\xi'^2)}\,d\xi\,d\xi',$$

since $u^2+u'^2$, the square of the distance from the origin, is equal to $\xi^2+\xi'^2$; and $du\,du'$, an element of area in the plane, may be replaced by $d\xi\,d\xi'$. Thus formula (136) for the number of collisions between molecules of classes A and B becomes

$$\pi\nu^2\left(\frac{hm}{\pi}\right)^3 e^{-hm(\xi^2+\eta^2+\zeta^2+\xi'^2+\eta'^2+\zeta'^2)}\,V\sigma^2\,d\xi\,d\eta\,d\zeta\,d\xi'\,d\eta'\,d\zeta',$$

$$\ldots\ldots(138)$$

in which $\eta = \frac{1}{\sqrt{2}}(v'-v)$, etc. The reason for this transformation of coordinates is seen as soon as we express the value of V in the new coordinates, for we find that

$$V^2 = (u'-u)^2 + (v'-v)^2 + (w'-w)^2$$
$$= 2(\xi^2+\eta^2+\zeta^2),$$

so that V does not depend on ξ', η', ζ'. Thus we may at once integrate formula (138) for all values of ξ', η' and ζ', obtaining

$$\pi\nu^2\left(\frac{hm}{\pi}\right)^{\frac{3}{2}} e^{-hm(\xi^2+\eta^2+\zeta^2)}\,V\sigma^2\,d\xi\,d\eta\,d\zeta.$$

This is not, as might at first be thought, the number of collisions in which ξ, η, ζ lie within limits $d\xi, d\eta, d\zeta$; it is twice that number, since each collision has been counted twice—once as a collision between a molecule of class A and one of class B, and again as a collision between a molecule of class B and one of class A. Thus the true number of collisions in which ξ, η, ζ lie within assigned limits $d\xi\, d\eta\, d\zeta$ is half the foregoing, or

$$\tfrac{1}{2}\pi \nu^2 \left(\frac{hm}{\pi}\right)^{\frac{3}{2}} e^{-hm(\xi^2+\eta^2+\zeta^2)} V\sigma^2 d\xi\, d\eta\, d\zeta. \qquad \ldots\ldots(139)$$

There is an obvious resemblance to Maxwell's law of distribution of velocities. Again we see that all directions for ξ, η, ζ are equally likely, and as the relative velocity V has components $\sqrt{2}\xi, \sqrt{2}\eta, \sqrt{2}\zeta$, it follows that all directions are equally likely for V, the relative velocity before collision. We have also just seen (§ 105) that, whatever the directions may be of the velocities and relative velocity of the molecules before collision, all directions are equally likely for the relative velocity after collision.

107. Integrating formula (139) over all possible directions for V, we obtain, as in § 106,

$$\nu^2\sigma^2 \sqrt{\frac{\pi h^3 m^3}{2}}\, e^{-\frac{1}{2}hmV^2} V^3 dV, \qquad \ldots\ldots(140)$$

as the total number of collisions in which the relative velocity lies between V and $V + dV$. If we finally integrate this from $V = 0$ to $V = \infty$, we obtain for the total number of collisions of all kinds

$$\nu^2\sigma^2 \sqrt{\frac{2\pi}{hm}}. \qquad \ldots\ldots(141)$$

Since (§ 91) $\bar{c} = 2(\pi hm)^{-\frac{1}{2}}$, this may be replaced by

$$\frac{\pi}{\sqrt{2}} \nu^2\sigma^2 \bar{c}.$$

These results will be required later.

Free Path

108. The free path is usually defined as the distance which a molecule travels between one collision and the next. We have seen that the ν molecules in a unit volume of gas experience

$$\frac{\pi}{\sqrt{2}}\,\nu^2\sigma^2\,\bar{c}$$

collisions per unit time. Since each collision terminates two free paths, the total number of free paths is twice the foregoing expression, or

$$\sqrt{2}\pi\nu^2\sigma^2\,\bar{c}.$$

The total length of all these free paths is of course the total distance travelled by the ν molecules in unit time, and this is $\nu\bar{c}$. Hence by division we find that the average length of a free path is

$$\frac{1}{\sqrt{2}\pi\nu\sigma^2} = \frac{0.7071}{\pi\nu\sigma^2}. \qquad \ldots\ldots(142)$$

This is generally described as Maxwell's mean free path; as we shall see later (§§ 117, 135), there are other ways of defining and evaluating the mean free path in a gas.

Free Path in a Mixture of Gases

109. It is rather more difficult to calculate the free path in a mixture of gases in which there are molecules of different sizes.

As before, the constants of the molecules of different types will be distinguished by suffixes, those of the first type having a suffix unity $(m_1, \sigma_1, \nu_1, f_1)$, and so on. We shall require a system of symbols to denote the distances apart at collision of the centres of two molecules of different kinds. Let these be S_{11}, S_{12}, S_{23}, etc., S_{pq} being the distance of the centres of two molecules of types p, q when in collision. Obviously

$$S_{12} = \tfrac{1}{2}(\sigma_1 + \sigma_2), \quad S_{11} = \sigma_1, \text{ etc.}$$

As in § 89, we find for the number of collisions per unit time between molecules of types 1 and 2 and of classes A and B respectively (as defined in § 85)

$$\nu_1\nu_2 f_1(u,v,w) f_2(u',v',w')\, V S_{12}^2 \cos\theta\, du\,dv\,dw\,du'\,dv'\,dw'\,d\omega,$$

where V is the relative velocity, and $d\omega$ the element of solid angle to within which the line of centres is limited.

Replacing f_1, f_2 by the values appropriate to the steady state, and carrying out the integration with respect to $d\omega$, the number of collisions per unit time is found to be

$$\pi \nu_1 \nu_2 \left(\frac{h^3 m_1^{\frac{3}{2}} m_2^{\frac{3}{2}}}{\pi^3} e^{-h(m_1 c^2 + m_2 c'^2)} \right) V S_{12}^2 \, du \, dv \, dw \, du' \, dv' \, dw',$$

$$\dots\dots(143)$$

this expression being exactly analogous to expression (136) previously obtained.

Let the velocity-components u, v, w, u', v', w' now be replaced by new variables, $\mathbf{u}, \mathbf{v}, \mathbf{w}, \alpha, \beta, \gamma$ given by

$$\mathbf{u} = \frac{m_1 u + m_2 u'}{m_1 + m_2}, \text{ etc.}; \quad \alpha = u' - u, \text{ etc.}, \quad \dots\dots(144)$$

so that $\mathbf{u}, \mathbf{v}, \mathbf{w}$ are the components of velocity of the centre of gravity of the two molecules, and α, β, γ are as usual the components of the relative velocity V. Writing

$$\mathbf{u}^2 + \mathbf{v}^2 + \mathbf{w}^2 = \mathbf{c}^2, \quad \alpha^2 + \beta^2 + \gamma^2 = V^2,$$

we find that

$$m_1 c^2 + m_2 c'^2 = (m_1 + m_2) \, \mathbf{c}^2 + \frac{m_1 m_2}{m_1 + m_2} V^2.$$

We readily find, by the method already used in § 106, that $du \, du' = d\mathbf{u} \, d\alpha$, so that expression (143) may be replaced by

$$\pi \nu_1 \nu_2 \left(\frac{h^3 m_1^{\frac{3}{2}} m_2^{\frac{3}{2}}}{\pi^3} e^{-h \left[(m_1 + m_2) \, \mathbf{c}^2 + \frac{m_1 m_2}{m_1 + m_2} V^2 \right]} \right) V S_{12}^2 \, d\mathbf{u} \, d\mathbf{v} \, d\mathbf{w} \, d\alpha \, d\beta \, d\gamma.$$

$$\dots\dots(145)$$

110. On integrating with respect to all possible directions in space for the velocity $\mathbf{c}$ of the centre of gravity, we may replace $d\mathbf{u} \, d\mathbf{v} \, d\mathbf{w}$ by $4\pi \mathbf{c}^2 \, d\mathbf{c}$ while similarly, integrating with respect to all possible directions for V, we may replace $d\alpha \, d\beta \, d\gamma$ by $4\pi V^2 dV$. The number of collisions per unit volume per unit time for which $\mathbf{c}, V$ lie within specified small ranges $d\mathbf{c} \, dV$ is therefore

$$16 \nu_1 \nu_2 h^3 m_1^{\frac{3}{2}} m_2^{\frac{3}{2}} S_{12}^2 e^{-h \left[(m_1 + m_2) \, \mathbf{c}^2 + \frac{m_1 m_2}{m_1 + m_2} V^2 \right]} \mathbf{c}^2 V^3 \, d\mathbf{c} \, dV.$$

$$\dots\dots(146)$$

Integrating from $c = 0$ to $c = \infty$, the number of collisions for which V lies between V and $V + dV$ is found to be

$$4\nu_1\nu_2 \sqrt{\frac{\pi h^3 m_1^3 m_2^3}{(m_1 + m_2)^3}} S_{12}^2 e^{-\frac{h m_1 m_2}{m_1 + m_2} V^2} V^3 dV, \qquad \ldots\ldots(147)$$

and again integrating this expression from $V = 0$ to $V = \infty$, the total number of collisions per unit volume per unit time between molecules of types 1 and 2 is found to be

$$2\nu_1\nu_2 S_{12}^2 \sqrt{\frac{\pi}{h}\left(\frac{1}{m_1} + \frac{1}{m_2}\right)}. \qquad \ldots\ldots(148)$$

This formula gives the number of free paths of the ν_1 molecules of the first type in unit volume, which are terminated per unit time by molecules of the second type. When the difference between the two types of molecules is ignored, it reduces to twice expression (141) already found, the reason for the multiplying factor 2 being that already explained on p. 134.

111. If we divide expression (148) by ν_1, we obtain the mean chance of collision per unit time for a molecule of type 1 with a molecule of type 2. Summing this over all types of molecule, the total mean chance of collision per unit time for a molecule of type 1 is found to be

$$2\Sigma\nu_s S_{1s}^2 \sqrt{\frac{\pi}{h}\left(\frac{1}{m_1} + \frac{1}{m_s}\right)}, \qquad \ldots\ldots(149)$$

where the summation is taken over all types of molecule. The mean time interval between collisions is of course the reciprocal of this.

The total distance described by ν_1 molecules of the first kind per unit time is

$$\nu_1 \bar{c}_1 = \frac{2\nu_1}{\sqrt{\pi h m_1}}, \qquad \ldots\ldots(150)$$

while the total number of free paths described by these ν_1 molecules is equal to ν_1 times expression (149). By division, the mean free path for molecules of the first type is found to be

$$\lambda_1 = \frac{1}{\pi\Sigma\nu_s S_{1s}^2 \sqrt{1 + \dfrac{m_1}{m_s}}}. \qquad \ldots\ldots(151)$$

When there is only one kind of gas present, this reduces to formula (140) already found.

If there are only two kinds of gas present, one with molecules of far greater mass than the other $(m_2 > m_1)$, the lighter molecules will move enormously more rapidly than the heavier because of the equipartition of energy. We may now put $m_1/m_2 = 0$, and formula (151) reduces to

$$\lambda_1 = \frac{1}{\sqrt{2}\pi\nu_1 S_{11}^2 + \pi\nu_2 S_{12}^2}.$$

If the diameter of the lighter molecules is much smaller than that of the more massive, the first term in the denominator may be neglected, and the free path is given by

$$\lambda_1 = \frac{1}{\pi\nu_2 S_{12}^2} = \frac{4}{\pi\nu_2 \sigma_2^2}. \qquad \ldots\ldots(152)$$

This is the simple case already considered in § 26, our formula (152) agreeing with the formula there obtained. It gives the free path of electrons threading their way through a gas.

DEPENDENCE OF FREE PATH ON VELOCITY

112. This last calculation will have suggested that the free path of a molecule may depend very appreciably on the velocity of the molecule; a molecule moving with exceptional speed may expect a free path of exceptional length, while obviously a molecule moving with zero velocity can expect only a free path of zero length.

It is important to examine the exact correlation between the speed of a molecule and its expectation of free path.

We shall suppose that the gas consists of a mixture of molecules of different types, as in § 109.

Let us fix our attention on a molecule of the first type, moving with velocity c. The chance of collision per unit time with a molecule of the second type having a specified velocity c' is equal to the probable number of molecules of this second kind in a cylinder of base πS_{12}^2, and of height V, where V is the relative velocity.

The second molecule is supposed to have a velocity c'. Let θ, ϕ be angles determining the direction of this velocity, θ measuring the angle between its direction and that of c, and ϕ being an azimuth. The number of molecules of the second kind per unit volume for which c', θ, ϕ lie within small specified ranges $dc', d\theta, d\phi$ is

$$\nu_2 \left(\frac{hm_2}{\pi} \right)^{\frac{3}{2}} e^{-hm_2 c'^2} c'^2 \sin \theta \, d\theta \, d\phi \, dc'.$$

The result of integrating with respect to ϕ is obtained by replacing $d\phi$ by 2π, and on multiplying this by $\pi S_{12}^2 V$, we obtain for the number of molecules of the second kind which lie within the cylinder of volume $\pi S_{12}^2 V$ and are such that c', θ lie within a range $dc', d\theta$,

$$2\nu_2 S_{12}^2 \sqrt{\pi h^3 m_2^3} \, V e^{-hm_2 c'^2} c'^2 \sin \theta \, d\theta \, dc'. \qquad \ldots\ldots(153)$$

When c, c' are kept constant, the value of V varies only with θ, being given by

$$V^2 = c^2 + c'^2 - 2cc' \cos \theta,$$

whence we obtain by differentiation, still keeping c, c' constant,

$$V dV = cc' \sin \theta \, d\theta.$$

Thus we can replace expression (153) by

$$2\nu_2 S_{12}^2 \sqrt{\pi h^3 m_2^3} \, e^{-hm_2 c'^2} \frac{c'}{c} \, dc' V^2 dV, \qquad \ldots\ldots(154)$$

and proceed to integrate with respect to V. The limits for V are $c + c'$ and $c \sim c'$, so that

$$\int V^2 dV = \tfrac{2}{3} c (c^2 + 3c'^2) \text{ when } c' > c$$
$$= \tfrac{2}{3} c' (c'^2 + 3c^2) \text{ when } c' < c.$$

Thus the result of integrating expression (154) with respect to V is

when $c' > c$, $\quad \tfrac{4}{3}\nu_2 S_{12}^2 \sqrt{\pi h^3 m_2^3} e^{-hm_2 c'^2} c' (c^2 + 3c'^2) \, dc', \qquad \ldots\ldots(155)$

when $c' < c$, $\quad \tfrac{4}{3}\nu_2 S_{12}^2 \sqrt{\pi h^3 m_2^3} e^{-hm_2 c'^2} \frac{c'^2}{c} (c'^2 + 3c^2) \, dc'. \qquad \ldots\ldots(156)$

If we now integrate this quantity with respect to c' from $c' = 0$ to $c' = \infty$ (using the appropriate form according as c' is greater or less than c), we obtain, as the aggregate chance per unit time of a collision between a given molecule of the first type moving with velocity c, and a molecule of the second type moving with velocity c',

$$\tfrac{4}{3}\nu_2 S_{12}^2 \sqrt{h\pi^3 m_2^3}\left[\int_c^\infty c'(c^2+3c'^2)\, e^{-hm_2c'^2}\, dc' \right.$$
$$\left. + \int_0^c \frac{c'^2(c'^2+3c^2)}{c}\, e^{-hm_2c'^2}\, dc'\right]. \quad \ldots\ldots(157)$$

The former of the two integrals inside the square bracket can be evaluated directly. To integrate the second integral, we replace $hm_2c'^2$ by y^2. After continued integration with respect to y^2, we find as the sum of the two integrals in expression (157),

$$\frac{3}{4c\sqrt{h^5 m_2^5}}\left[c\sqrt{hm_2}\, e^{-hm_2c^2} + (2hm_2c^2+1)\int_0^{c\sqrt{hm_2}} e^{-y^2}\, dy\right].$$
$$\ldots\ldots(158)$$

If we introduce a function* $\psi(x)$ defined by

$$\psi(x) = xe^{-x^2} + (2x^2+1)\int_0^x e^{-y^2}\, dy, \quad \ldots\ldots(159)$$

expression (158) may be expressed in the form

$$\frac{3}{4c\sqrt{h^5 m_2^5}}\, \psi(c\sqrt{hm_2}),$$

and hence if we denote expression (157) by Θ_{12}, its value is found to be

$$\Theta_{12} = \frac{\sqrt{\pi}\,\nu_2 S_{12}^2}{hm_2c}\, \psi(c\sqrt{hm_2}). \quad \ldots\ldots(160)$$

With this definition of Θ_{12} we see that when a molecule of the first kind is moving with a velocity c, the chance that it collides with a molecule of the second kind in time dt is $\Theta_{12}dt$.

113. If we change the suffix 2 into 1 wherever it occurs, we obtain an expression Θ_{11} for the chance per unit time that a molecule of the first kind moving with velocity c shall collide with another molecule of the same kind.

---

* The value of $\int_0^x e^{-y^2}\, dy$ cannot be expressed in simpler terms, so that $\psi(x)$ as defined by equation (159) is already in its simplest form. Tables for the evaluation of $\psi(x)$ are given in Appendix v.

By addition, the total chance per unit time that a molecule of the first kind moving with velocity c shall collide with a molecule of any kind is

$$\Sigma\Theta_{1s} = \Theta_{11} + \Theta_{12} + \Theta_{13} + \dots \qquad \dots\dots(161)$$

In unit time the molecule we are considering describes a distance c; hence the chance of collision per unit length of path is

$$\frac{1}{c}\Sigma\Theta_{1s}. \qquad \dots\dots(162)$$

The mean free path λ_c, for molecules of the first kind moving with velocity c, is accordingly

$$\lambda_c = \frac{c}{\Sigma\Theta_{1s}}. \qquad \dots\dots(163)$$

When there is only one kind of gas, this assumes the form

$$\lambda_c = \frac{c}{\Theta} = \frac{hmc^2}{\sqrt{\pi}\nu\sigma^2\psi(c\sqrt{hm})}. \qquad \dots\dots(164)$$

114. This formula expressing λ_c as a function of c is, unhappily, too complex to convey much definite meaning to the mind, and we are therefore compelled to fall back on numerical values. The following table, which is taken from Meyer's *Kinetic Theory of Gases* (p. 429), gives the ratio of λ_c (equation (164)) to Maxwell's mean free path λ (equation (142)) for different values of c, from $c = 0$ to $c = \infty$.

$c/\bar{c}$	hmc^2	λ_c/λ	λ/λ_c
0	–	0	∞
0·25	–	0·3445	2·9112
0·5	–	0·6411	1·5604
0·627	$\frac{1}{2}$	0·7647	1·3111
0·886	1	0·9611	1·0407
1·0	–	1·0257	0·9749
1·253	2	1·1340	0·8819
1·535	3	1·2127	0·8247
1·772	4	1·2572	0·7954
2	–	1·2878	0·7765
3	–	1·3551	0·7380
4	–	1·3803	0·7244
5	–	1·3923	0·7182
6	–	1·3989	0·7149
∞	–	1·4142	0·7071

We notice that for infinite velocity $\lambda_c = \sqrt{2}\lambda$, as previously found.

Probability of a Free Path of given Length

115. It is of interest to find the probability that a molecule shall describe a free path of any assigned length.

Let $f(l)$ denote the probability that a molecule moving with a velocity c shall describe a free path at least equal to l. After the molecule has described a distance l, the chance of collision within a further distance dl is, by formula (162), equal to dl/λ_c. Hence the chance that a molecule shall describe a distance l, and then a further distance dl, without collision is

$$f(l)\,(1 - dl/\lambda_c).$$

This must however be the same thing as $f(l + dl)$ or

$$f(l) + \frac{\partial f(l)}{\partial l}\,dl.$$

Equating these expressions, we have

$$\frac{\partial f(l)}{\partial l} = -\frac{f(l)}{\lambda_c},$$

of which the solution is

$$f(l) = e^{-l/\lambda_c}, \qquad\qquad \text{......(165)}$$

the arbitrary constant of the integration being determined by the condition that $f(0) = 1$. Thus a shower of N molecules, each originally moving with the same velocity c, will be reduced in number to

$$Ne^{-l/\lambda_c} \qquad\qquad \text{......(166)}$$

after travelling a distance l through the gas. Typical numerical values are given below, N being taken to be 100:

$\frac{l}{\lambda_c} =$	0	0·01	0·02	0·1	0·2	0·25	0·33	0·50	1	2	3	4	4·16
$Ne^{-l/\lambda_c} =$	100	99	98	90	82	78	72	61	37	14	5	2	1

The occurrence of the exponential in expression (166) shews that free paths which are many times greater than the mean free path will be extremely rare. For instance, only one molecule in 148 will describe a path as great as $5\lambda_c$, only one in 22,027 a path as great as $10\lambda_c$, only one in $2\cdot7 \times 10^{43}$ a path as great as $100\lambda_c$, and so on.

116. Formula (166) can be tested experimentally in various ways, and each test provides a value of the free path λ_c, from which it is then possible to deduce the molecular diameter σ.

When a shower of electrons or other swiftly moving particles is shot through a gas, it is natural to assume a law of diminution of the form

$$i = i_0 e^{-\alpha x}, \qquad\qquad ......(167)$$

where i_0 is the initial current, i the current after a distance x, and α is a coefficient of extinction. Innumerable experimenters have found that a law of this type represents the facts. We see that formula (167) is identical with (166), α being the reciprocal of the free path λ_c.

In recent years, a number of experimenters* have tested formulae of this type for beams of electrons projected into gases with low velocities which are comparable with those which occur in the kinetic theory. In general the lengths of free path and molecular diameters deduced from them agree well with those calculated in other ways. Here and there, however, difficult questions arise. The most difficult is perhaps connected with electrons of low velocity moving through the monatomic gases— helium, argon, neon, krypton and xenon. As a certain electronic speed is approached, the mean free path undergoes a violent decrease, as though the atomic diameter was greatly increased. For still lower velocities the free path is abnormally long, almost as though the atoms had become transparent to the electrons. The kinetic theory alone is not able to provide a satisfactory explanation of these phenomena; to obtain this we must avail ourselves of the methods of wave-mechanics. But until we come to these quite low electronic velocities, experiment shews that collisions occur in the way predicted by pure kinetic theory.

Max Born[†] has tested formula (167) for the free path of complete atoms with satisfactory results. A beam of silver atoms is projected in the usual way from an electric furnace, so that each

---

* H. F. Mayer, *Ann. d. Phys.* **64** (1921), p. 451. C. Ramsauer, *Ann. d. Phys.* **64** (1921), p. 513. R. B. Brode, *Phys. Rev.* **25** (1925), p. 636. M. Rusch, *Ann. d. Phys.* **80** (1926), p. 707. R. Kollath, *Phys. Zeitschr.* **31** (1930), p. 986.

† *Phys. Zeitschr.* **21** (1920), p. 578.

atom of the beam falls on one or other of four screens of glass cooled to liquid air temperature. These four screens are at different distances x_1, x_2, x_3, x_4. The number of molecules which strike the screens per unit area can be estimated from the density of silver deposit on the glass, and are found to be approximately in the ratio

$$e^{-\alpha x_1} : e^{-\alpha x_2} : e^{-\alpha x_3} : e^{-\alpha x_4}.$$

In this way, Born found the free path of silver atoms moving through air at atmospheric pressure to be about $1 \cdot 3 \times 10^{-5}$ cm. More recently Bielz,* using a refinement of the same method, has found the free path of silver atoms moving through nitrogen at atmospheric pressure to be about $1 \cdot 29 \times 10^{-5}$ cm.

F. Knauer† has tested the formula for the scattering of a beam of complete molecules, shot both into the same gas as the beam and also into mercury vapour. The predictions of the kinetic theory are confirmed at high velocities, but at low velocities it is found necessary to call in the wave-mechanics to explain the observations.

117. On differentiating formula (165), we see that the probability that a molecule moving with a velocity c shall describe a free path of length between l and $l + dl$ is

$$e^{-l/\lambda_c} \frac{dl}{\lambda_c}. \qquad \qquad \dots\dots(168)$$

The foregoing results apply only to molecules moving with a given velocity c. If the velocities are distributed in accordance with Maxwell's law, then the fraction of the whole number of molecules which at any given instant have described a distance greater than l since their last collision will be

$$\sqrt{\frac{\pi^3}{h^3 m^3}} \int_0^\infty 4\pi c^2 e^{-hmc^2 - l/\lambda_c} dc. \qquad \dots\dots(169)$$

This function is not easy to calculate in any way. As the result of a rough calculation by quadrature, I have found that through the range of values for l in which its value is appreciable, it does

---

* *Zeitschr. f. Phys.* **32** (1925), p. 81.
† *Zeitschr. f. Phys.* **80** (1933), p. 80 and **90** (1934), p. 559.

not ever differ by more than about 1 per cent. from $e^{-1.04l/\lambda}$, which is the value for molecules moving with velocity $1/\sqrt{hm}$.

At any instant a fraction

$$\sqrt{\frac{h^3m^3}{\pi^3}}\, 4\pi e^{-hmc^2} c^2 dc$$

of the whole number of molecules is moving with velocity c. These molecules, on the average, are starting to describe distances c/Θ each before collision. Hence the average distance λ_T that all the molecules will travel before collision is given by

$$\lambda_T = \int_0^\infty \frac{c}{\Theta}\sqrt{\frac{h^3m^3}{\pi^3}}\, 4\pi e^{-hmc^2} c^2 dc = \frac{1}{\pi v \sigma^2}\int_0^\infty \frac{4x^4 e^{-x^2} dx}{\psi(x)}.$$

This integral can only be evaluated by quadrature. The evaluation has been performed by Tait[*] and Boltzmann,[†] who agree in assigning to it the value 0·677, leading to a value for λ_T which is some 4 per cent less than the value of the free path calculated in § 108.

It must be noticed that these two free paths are calculated in different ways. That calculated in § 108 is the mean of all the paths described in unit time, that just calculated is the mean of all the paths being described at a given instant, or to be more precise, is the mean of all the distances described from a given instant to the next collision (cf. § 134 below). This latter is commonly known as Tait's free path.

LAW OF DISTRIBUTION OF VELOCITIES IN COLLISION

118. In many physical problems, it is important to consider the distribution of relative velocities, ratios of velocities, etc. in the different collisions which occur. We attempt to obtain expressions for various laws of distribution of this type.

In formulae (155) and (156) we obtained expressions for the chance per unit time that a molecule of type 1 moving with

---

[*] *Edinburgh Trans.* **33** (1886), p. 74.
[†] *Wien. Sitzungsber.* **96** (1887), p. 905, of *Gastheorie*, **1**, p. 73.

velocity c should collide with a molecule of type 2 moving with a velocity between c' and $c' + dc'$.

The number of molecules of type 1 per unit volume moving with a velocity between c and $c + dc$ is

$$4\pi\nu_1\left(\frac{hm}{\pi}\right)^{\frac{3}{2}} e^{-hm_1c^2} c^2 dc.$$

Multiplying expressions (155) and (156) by this number we obtain, as the total number of collisions per unit time per unit volume between molecules having specified velocities within ranges dc, dc',

when $c' > c$,

$$\tfrac{16}{3}\nu_1\nu_2 S_{12}^2 h^3 m_1^{\frac{3}{2}} m_2^{\frac{3}{2}} e^{-h(m_1c^2 + m_2c'^2)} c^2 c'(c^2 + 3c'^2)\, dc\, dc', \quad ......(170)$$

when $c' < c$,

$$\tfrac{16}{3}\nu_1\nu_2 S_{12}^2 h^3 m_1^{\frac{3}{2}} m_2^{\frac{3}{2}} e^{-h(m_1c^2 + m_2c'^2)} cc'^2(3c^2 + c'^2)\, dc\, dc'. \quad ......(171)$$

119. We proceed next to find the number of collisions in which the velocities c, c' stand in a given ratio to one another. Let $c = \kappa c'$, and let the variables in expressions (170) and (171) be changed from c, c' to κ, c'. Clearly the differential $dc\, dc'$ becomes $c' d\kappa\, dc'$, and the two expressions become

when $\kappa > 1$,

$$\tfrac{16}{3}\nu_1\nu_2 S_{12}^2 h^3 m_1^{\frac{3}{2}} m_2^{\frac{3}{2}} e^{-hc'^2(m_1\kappa^2 + m_2)} \kappa(3\kappa^2 + 1)\, d\kappa c'^6 dc', \quad ......(172)$$

when $\kappa < 1$,

$$\tfrac{16}{3}\nu_1\nu_2 S_{12}^2 h^3 m_1^{\frac{3}{2}} m_2^{\frac{3}{2}} e^{-hc'^2(m_1\kappa^2 + m_2)} \kappa^2(\kappa^2 + 3)\, d\kappa c'^6 dc'. \quad ......(173)$$

On integrating these expressions with respect to c' from $c' = 0$ to $c' = \infty$, we obtain the number of collisions for which κ, the ratio of the velocities, lies within a given range $d\kappa$. The numbers are readily found to be

when $\kappa > 1$, $$5\nu_1\nu_2 S_{12}^2\left(\frac{\pi m_1^3 m_2^3}{h}\right)^{\frac{1}{2}} \frac{\kappa(3\kappa^2 + 1)}{(m_1\kappa^2 + m_2)^{\frac{7}{2}}} d\kappa, \quad ......(174)$$

when $\kappa < 1$, $$5\nu_1\nu_2 S_{12}^2\left(\frac{\pi m_1^3 m_2^3}{h}\right)^{\frac{1}{2}} \frac{\kappa^2(\kappa^2 + 3)}{(m_1\kappa^2 + m_2)^{\frac{7}{2}}} d\kappa. \quad ......(175)$$

The total number of collisions per unit time per unit volume may of course be derived by integrating this quantity from $\kappa = 0$ to $\kappa = \infty$. It is found to be

$$2\nu_1\nu_2 S_{12}^2 \sqrt{\frac{\pi}{h}\left(\frac{1}{m_1}+\frac{1}{m_2}\right)}, \qquad \ldots\ldots(176)$$

which agrees, as it ought, with formula (148).

120. The law of distribution of κ in different collisions can be obtained by dividing expressions (174) and (175) by the total number of collisions (176). This law of distribution is found to be

when $\kappa > 1$,

$$\frac{5}{2}\frac{m_1^2 m_2^2}{(m_1+m_2)^{\frac{1}{2}}}\frac{\kappa(3\kappa^2+1)}{(m_1\kappa^2+m_2)^{\frac{7}{2}}}d\kappa, \qquad \ldots\ldots(177)$$

when $\kappa < 1$,

$$\frac{5}{2}\frac{m_1^2 m_2^2}{(m_1+m_2)^{\frac{1}{2}}}\frac{\kappa^2(\kappa^2+3)}{(m_1\kappa^2+m_2)^{\frac{7}{2}}}d\kappa. \qquad \ldots\ldots(178)$$

The law of distribution of values of κ when the molecules are similar is obtained on taking $m_1 = m_2$. We must notice however that if we simply put $m_1 = m_2$ in expressions (177) and (178) each collision is counted twice, once as having a ratio of velocities κ and once as having a ratio of velocities $1/\kappa$. It seems simplest to define the value of κ for a collision in this case as the ratio of the greater to the smaller velocity, so that κ is always greater than unity, and we then obtain the law of distribution by putting $m_1 = m_2$ in expression (177) and multiplying by two so that each collision shall only count once. The law of distribution is found to be

$$\frac{5\kappa(3\kappa^2+1)}{\sqrt{2}(1+\kappa^2)^{\frac{7}{2}}}d\kappa, \qquad \ldots\ldots(179)$$

of which the value when integrated from $\kappa = 1$ to $\kappa = \infty$ is unity, as it ought to be.

PERSISTENCE OF VELOCITY AFTER COLLISION

121. The next problem will be to examine the average effect of a collision as regards reversal or deflection of path. We shall find that in general a collision does not necessarily reverse the velocity in the original direction of motion, or even reduce it to rest: there is a marked tendency for the original velocity to persist to some extent after collision. It is obviously of the utmost importance to form an estimate of the extent to which this persistence of velocity occurs.

Persistence of Velocity when the Molecules are Similar Elastic Spheres

122. Let us again represent molecules by elastic spheres, and consider two molecules of equal mass colliding with velocities c, c'. In fig. 29 let OP and OQ represent these velocities, and let R be the middle point of PQ. Then we can resolve the motion of the two molecules into

(i) a motion of the centre of mass of the two, the velocity of this motion being represented by OR, and

(ii) two equal and opposite velocities relative to the centre of mass, these being represented by RP and RQ.

Imagine a plane RTS drawn through R parallel to the common tangent to the spheres at the moment of impact, and let P', Q'

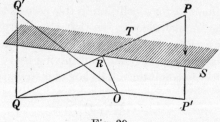

Fig. 29

be the images of P, Q in this plane. Then clearly RP' and RQ' represent the velocities relatively to the centre of gravity after impact, so that OP' and OQ' represent the actual velocities in space.

We have already seen (§ 104) that all directions are equally likely for RP', RQ', the velocities after collision, so that the "expectation" of the component of velocity of either molecule after impact in any direction is equal to the component of OR in that direction.

123. Let us now average over all possible directions for the velocity of the second molecule, keeping the magnitude of this velocity constant. In fig. 30 let OP, OQ as before represent the velocities of the two colliding molecules, and let R be the middle point of PQ, so that OR represents the velocity of the centre of gravity of the two molecules. We have to average the components of the velocity OR over all positions of Q which lie on a sphere having O for centre. It is at once obvious that the average component of OR in any direction perpendicular to OP is zero. We have, therefore, only to find the component in the direction OP, say ON. We must not suppose all directions for OQ to be equally likely, for (cf. § 85) the probability of collision with any two

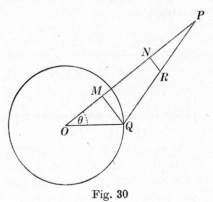

Fig. 30

velocities is proportional to the relative velocity. Thus the probability of the angle POQ lying between θ and $\theta + d\theta$ is not simply proportional to $\sin\theta\, d\theta$, but is proportional to $PQ\sin\theta\, d\theta$, for PQ represents the relative velocity. The average value of the component ON is therefore

$$\overline{ON} = \frac{\int_0^\pi ON \cdot PQ \sin\theta\, d\theta}{\int_0^\pi PQ \sin\theta\, d\theta}. \qquad \ldots\ldots(180)$$

Let us now write

$$OP = c, \quad OQ = c', \quad PQ = V,$$

so that

$$V^2 = c^2 + c'^2 - 2cc'\cos\theta. \qquad \ldots\ldots(181)$$

Then

$$ON = \tfrac{1}{2}(OP + OM) = \tfrac{1}{2}(c + c' \cos\theta) = \frac{1}{4c}(3c^2 + c'^2 - V^2).$$

$$......(182)$$

By differentiation of relation (181), we have

$$V \, dV = cc' \sin\theta \, d\theta,$$

so that equation (180) becomes

$$\overline{ON} = \frac{\int (3c^2 + c'^2 - V^2) V^2 dV}{4c \int V^2 dV} = \frac{3c^2 + c'^2}{4c} - \frac{\int V^4 dV}{4c \int V^2 dV},$$

$$......(183)$$

the limits of integration being from $V = c' \sim c$ to $V = c' + c$.

Performing the integration, we find that

when $c > c'$,

$$\overline{ON} = \frac{15c^4 + c'^4}{10c(3c^2 + c'^2)}, \qquad ......(184)$$

when $c < c'$,

$$\overline{ON} = \frac{c(5c'^2 + 3c^2)}{5(3c'^2 + c^2)}. \qquad ......(185)$$

124. Since these expressions are positive for all values of c and c', we see that whatever the velocities of the two colliding molecules may be, the "expectation" of the velocity of the first molecule after collision is definitely in the same direction as the velocity before collision. Naturally the same also is true of the second molecule.

If we denote $\overline{ON}$, the "expectation" of velocity after collision of the first molecule in the direction of OP, by α, then the ratio α/c may be regarded as a measure of the persistence of the velocity of the first molecule.

Formulae (184) and (185) give the values of α, and hence of the persistence α/c. It is at once seen that the values of α/c depend only on the ratio c/c', and not on the values of c and c' separately. If, as before (§ 119), we denote c/c' by κ, the values of the persistence are

when $\kappa > 1$,

$$\frac{\alpha}{c} = \frac{15\kappa^4 + 1}{10\kappa^2(3\kappa^2 + 1)}, \qquad ......(186)$$

when $\kappa < 1$,

$$\frac{\alpha}{c} = \frac{3\kappa^2 + 5}{5(\kappa^2 + 3)}. \qquad ......(187)$$

125. These expressions are too intricate to convey much meaning as they stand. The following table gives numerical values of the persistence α/c corresponding to different values of κ, the ratio of velocities:

$\dfrac{c}{c'} =$	∞	4	2	$1\frac{1}{2}$	1	$\frac{2}{3}$	$\frac{1}{2}$	$\frac{1}{4}$	0
$\dfrac{\alpha}{c} =$	0·500	0·492	0·473	0·441	0·400	0·368	0·354	0·339	0·333
$\dfrac{c'}{c} =$	0	$\frac{1}{4}$	$\frac{1}{2}$	$\frac{2}{3}$	1	$1\frac{1}{2}$	2	4	∞

It now appears that the persistence is a fraction which varies from $33\frac{1}{3}$ to 50 per cent, according to the ratio of the original velocities. From the values given, it is clear that we are likely to obtain fairly accurate results if we assume, for purposes of rough approximation, that the persistence is always equal to 40 per cent of the original velocity.

126. By averaging over all possible values of the ratio κ, we can obtain an exact value for the mean persistence averaged over all collisions.

Each collision involves two molecules of which the roles are entirely interchangeable. Let us agree to speak of the molecule of which the initial velocity is the greater as the first molecule, so that c/c' or κ is always greater than unity.

The persistences of the velocities of the two molecules involved in any one collision are respectively

$$\frac{15\kappa^4 + 1}{10\kappa^2(3\kappa^2 + 1)} \quad \text{and} \quad \frac{3 + 5\kappa^2}{5(3\kappa^2 + 1)},$$

the first of these expressions being given directly by formula (186), while the second is immediately obtained by writing $1/\kappa$ for κ in expression (187).

The mean persistence of the two molecules concerned in this collision, being the mean of the two expressions just found, is

$$\frac{25\kappa^4 + 6\kappa^2 + 1}{20\kappa^2(3\kappa^2 + 1)}. \qquad \ldots\ldots(188)$$

A few numerical values of this quantity are found to be:

$$\kappa = \quad 1 \quad\quad 1\tfrac{1}{4} \quad\quad 1\tfrac{1}{2} \quad\quad 2 \quad\quad 3 \quad\quad 4 \quad\quad \infty$$

mean persistence $= 0\cdot400 \quad 0\cdot401 \quad 0\cdot404 \quad 0\cdot413 \quad 0\cdot415 \quad 0\cdot416 \quad 0\cdot417$

The law of distribution of values of κ in the different collisions which occur has been found in formula (179) to be

$$\frac{5\kappa(3\kappa^2+1)}{\sqrt{2}(1+\kappa^2)^{\frac{7}{2}}} \, d\kappa. \qquad\qquad \ldots\ldots(189)$$

Multiplying together expressions (188) and (189) and integrating from $\kappa = 1$ to $\kappa = \infty$, we find that the mean persistence of all velocities after collision is

$$\frac{1}{4} + \frac{1}{4\sqrt{2}} \log_e(1+\sqrt{2}) = 0\cdot406.$$

Thus the average value of the persistence is very nearly equal to $\tfrac{2}{5}$, the value when the molecules collide with exactly equal velocities.

Persistence when Molecules have Different Masses

127. The calculations just given apply only when the molecules are all similar. Let us now examine what value is to be expected for the persistence when the molecules have different masses and sizes, but are still supposed to be elastic spheres.

Consider a collision between two molecules of masses m_1, m_2; let their velocities be a, b respectively as before, and let their relative velocity be V. Maxwell's result (§ 104) is still true that all directions are equally likely for the velocities after impact relative to the centre of gravity, and the expectation of any component of velocity after collision is exactly that of the common centre of gravity.

We may accordingly proceed to average exactly as in § 123. But if OR in fig. 30 represents the velocity of the centre of gravity, R will no longer be the middle point of PQ; it will divide PQ in such a way that $m_1 RP = m_2 RQ$. We then find, in place of relations (182),

$$ON = \frac{m_1 - m_2}{m_1 + m_2} c + \frac{m_2}{m_1 + m_2} (c + c' \cos \theta). \quad \ldots\ldots(190)$$

So long as c and c' are kept constant, we can average this exactly as before. The first term, being constant, is not affected by averaging, while the average value of the second term

$$\frac{m^2}{m_1 + m_2} (c + c' \cos \theta)$$

is equal to $\dfrac{2m_2}{m_1 + m_2}$ times the average of the term $\frac{1}{2}(c + c' \cos \theta)$ already found in § 123.

Hence, if $(\alpha/c)_e$ denotes the value of the persistence α/c when the two masses are equal, we have, in the general case in which the masses are unequal,

$$\frac{\alpha}{c} = \frac{m_1 - m_2}{m_1 + m_2} + \frac{2m_2}{m_1 + m_2} \left(\frac{\alpha}{c}\right)_e. \quad \ldots\ldots(191)$$

This gives the persistence of the velocity c of the molecule of mass m_1, the values of $(\alpha/c)_e$ being given by the table on p. 151. The persistence is of course a function of the two quantities m_1/m_2 and c'/c.

128. If we assume as a rough approximation that the value of $(\alpha/c)_e$ is equal to 0·400 regardless of the ratio of velocities κ, then equation (191) reduces to the approximate formula

$$\frac{\alpha}{c} = \frac{m_1 - \frac{1}{5}m_2}{m_1 + m_2}, \quad \ldots\ldots(192)$$

which of course depends only on m_1/m_2. This formula, however, must not be applied when the ratio m_1/m_2 is either very large or very small.

When m_1/m_2 is very small, c will be large compared with c' in practically all collisions, so that κ is very great and the appropriate value to assume for $(\alpha/c)_e$ is $\frac{1}{2}$, this corresponding to $\kappa = \infty$. From equation (191) we now obtain the approximate formula

$$\frac{\alpha}{c} = \frac{m_1}{m_1 + m_2} \ldots \left(\frac{m_1}{m_2} \text{ small}\right). \quad \ldots\ldots(193)$$

In the limit when m_1 vanishes in comparison with m_2, the persistence vanishes, as of course it must; at a collision the light molecule simply bounces off the heavy molecule; all directions can be seen to be equally likely by the method of § 104, and therefore the persistence is *nil*.

At the opposite extreme, when $\dfrac{m_1}{m_2}$ is very large, the appropriate values to assume are $\kappa = 0$ and $\left(\dfrac{\alpha}{c}\right)_e = \tfrac{1}{3}$. The approximate formula derived from equation (191) is now

$$\frac{\alpha}{c} = \frac{m_1 - \tfrac{1}{3}m_2}{m_1 + m_2} \dots \left(\frac{m_1}{m_2} \text{ large}\right). \qquad \dots\dots(194)$$

In the limit when m_2 vanishes, the persistence becomes equal to unity. This also can be seen directly: the heavy molecule merely knocks the light molecule out of its way, and passes on with its velocity unaltered.

129. In place of these approximate formulae, it is quite feasible to obtain a formula accurate for all values of m_1/m_2 by averaging the exact equation (191).

The result is

$$\overline{\left(\frac{\alpha}{c}\right)_e} = \frac{1}{4\mu^3\sqrt{1+\mu^2}} \log\left(\sqrt{1+\mu^2} + \mu\right) + \frac{1}{2} - \frac{1}{4\mu^2},$$

$$\dots\dots(195)$$

where $\mu^2 = m_2/m_1$. On substituting this into equation (191), we obtain a formula giving the average persistence of velocity for any ratio of masses.

130. It is readily verified that when $\mu = 1$ the value of this average persistence is 0.406, as already found in § 126.

When μ is very small, formula (195) reduces to

$$\left(\frac{\alpha}{c}\right)_e = \frac{1}{3} + \frac{2}{15}\mu^2 - \frac{4}{35}\mu^4 + \text{terms in } \mu^6, \text{ etc.,}$$

and similarly when μ is large, the expansion is

$$\left(\frac{\alpha}{c}\right)_e = \frac{1}{2} - \frac{1}{4\mu^2} + \frac{\log_e \mu}{4\mu^4} + \text{terms in } \frac{1}{\mu^6}, \text{ etc.}$$

From formulae (195) and (191), the following values can be calculated:

$$\frac{m_2}{m_1} = \quad 0 \quad \tfrac{1}{10} \quad \tfrac{1}{5} \quad \tfrac{1}{2} \quad 1 \quad 2 \quad 5 \quad 10 \quad \infty$$

$$\left(\overline{\frac{\alpha}{c}}\right)_e = 0\text{·}333 \quad 0\text{·}335 \quad 0\text{·}339 \quad 0\text{·}360 \quad 0\text{·}406 \quad 0\text{·}432 \quad 0\text{·}491 \quad 0\text{·}498 \quad 0\text{·}500$$

$$\left(\overline{\frac{\alpha}{c}}\right) = 1\text{·}000 \quad 0\text{·}879 \quad 0\text{·}779 \quad 0\text{·}573 \quad 0\text{·}406 \quad 0\text{·}243 \quad 0\text{·}152 \quad 0\text{·}086 \quad 0\text{·}000$$

These figures shew that the persistence is always positive but may have any value whatever, according to the ratio of the masses of the molecules.

131. For laws of force between molecules different from that between elastic spheres, the persistence of velocity will obviously be different from what it is for elastic spheres. Clearly, however, everything will depend on our definition of a collision. If we suppose that a very slight interaction is sufficient to constitute a collision, then the mean free path will be very short, while the persistence will be nearly equal to unity. If, on the other hand, we require large forces to come into play before calling a meeting of two molecules a collision, then the free path will be long, but the persistence will be small, or possibly even negative. In fact, the variations in the persistence of velocities just balance the arbitrariness of the standard we set up in defining a collision. This being so, it will be understood that the conception of persistence of velocities is hardly suited for use in cases where a collision is not a clearly defined event.

Chapter VI

VISCOSITY

132. At a collision between two molecules, energy, momentum and mass are all conserved. Energy, for instance, is neither created nor destroyed; a certain amount is transferred from one of the colliding molecules to the other. Thus the moving molecules may be regarded as transporters of energy, which they may hand on to other molecules when they collide with them. As the result of a long chain of collisions, energy may be transported from a region where the molecules have much energy to one where they have but little energy: studying such a chain of collisions we have in effect been studying the conduction of heat in a gas. If we examine the transport of momentum we shall find that we have been studying the viscosity of a gas—the subject of the present chapter. For viscosity represents a tendency for two contiguous layers of fluid to assume the same velocity, and this is effected by a transport of momentum from one layer to the other. Finally if we examine the transfer of the molecules themselves we study diffusion.

For the moment. we must study the transport of momentum. We think of the traversing of a free path of length λ as the transport of a certain amount of momentum through a distance λ. If the gas were in a steady state, every such transport would be exactly balanced by an equal and opposite transport in the reverse direction, so that the net transport would always be *nil*. But if the gas is not in a steady state, there will be an unbalanced residue, and this results in the phenomenon we wish to study.

General Equations of Viscosity

133. Consider a gas which is streaming in a direction parallel to the axis of x at every point as in fig. 31, the mass-velocity u_0 being different for different values of z, so that the streaming is in layers parallel to the plane of xy.

Let us denote mu, the x-momentum of any molecule by μ, and let $\bar{\mu}$ denote the mean value of μ at any point of the gas. Then $\bar{\mu}$ is equal to mu_0, and is a function of z only.

For definiteness let us suppose that z and $\bar{\mu}$ both increase as we move upwards in fig. 31. Molecules will be crossing any plane $z = z_0$ in both directions, and transporting a certain amount of μ across this plane as they do so. The amount of μ which any particular molecule transports across this plane will of course depend on the whole history of the molecule before meeting the plane. We shall however begin by supposing that it depends only on the history of the molecule since its last collision; we shall assume that the average molecule carries the amount of μ appropriate to the point at which this last collision occurred.

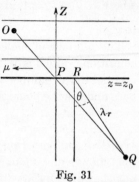

Fig. 31

134. Consider, then, a molecule which meets the plane $z = z_0$ at P, having previously come from a collision at Q. Let the velocity components of the molecule be u, v, w, and let the velocity be regarded as consisting of two parts:

(i) a velocity u_0, of components $u_0, 0, 0$, equal to the mass-velocity of the gas at P;

(ii) a velocity c, of components $u - u_0, v, w$, the molecular-velocity of the molecule relatively to the gas at P.

Let QP in fig. 31 represent the path of the molecule since its last collision, and let RP represent the distance travelled by the gas at P in the same interval of time, owing to its mass-velocity $u_0, 0, 0$. Then QR will represent the path described by the molecule relative to the mass-motion of the surrounding gas. Let the length QR be denoted by λ_r, and let this make an angle θ with the axis of z.

Since all directions of this molecular-velocity may be regarded as equally probable, the probability of θ lying between θ and $\theta + d\theta$ is proportional to $\sin \theta \, d\theta$. The number of molecules per unit volume having relative molecular-velocities for which

c and θ lie within specified small ranges dc, $d\theta$ may therefore be taken to be

$$\tfrac{1}{2}\nu f(c)\sin\theta\, d\theta\, dc, \qquad \dots\dots(196)$$

where $\displaystyle\int_0^\infty f(c)\, dc = 1$, in order that the total number may be equal to ν.

The number of molecules having a velocity satisfying these conditions, which cross a unit area of the plane $z = z_0$ in time dt, is equal to the number which at any instant occupy a cylinder of base unity in the plane $z = z_0$ and of height $c\cos\theta\, dt$; it is therefore

$$\tfrac{1}{2}\nu c f(c)\cos\theta\sin\theta\, d\theta\, dc\, dt. \qquad \dots\dots(197)$$

We shall suppose that on the average these molecules carry the amount of x-momentum μ appropriate to the point Q, and not that appropriate to the point P.

The z-co-ordinate of Q is

$$z_0 - \lambda_r\cos\theta,$$

so that the average value of μ at Q is less than that at P by an amount

$$\lambda_r\cos\theta\left(\frac{\partial\overline{\mu}}{\partial z}\right).$$

Hence the group of molecules under discussion carry, on the average, an amount of μ

$$-\lambda_c\cos\theta\left(\frac{\partial\overline{\mu}}{\partial z}\right) \qquad \dots\dots(198)$$

in excess of that appropriate to the point P, where λ_c is the mean free path for a molecule moving with velocity c.

The total excess of μ which these molecules carry across the plane $z = z_0$ is equal to the product of expressions (197) and (198), namely

$$-\tfrac{1}{2}\lambda_c\cos\theta\,\nu c f(c)\cos\theta\sin\theta\, d\theta\, dt.$$

On integrating the expression just found with respect to θ, we obtain the total transfer of momentum by all molecules with velocities between c and $c+dc$, whatever their direction. The limits for θ are 0 to π, values of θ from 0 to $\tfrac{1}{2}\pi$ covering molecules which cross the plane from below, and values of θ from $\tfrac{1}{2}\pi$ to π

those which cross the plane from above. The result of this integration is

$$\tfrac{1}{3}\nu c f(c) \lambda_c \left(\frac{\partial \overline{\mu}}{\partial z}\right) dc\, dt,$$

a negative sign indicating that the transfer is from above to below. On further integrating from $c = 0$ to $c = \infty$, we obtain for the amount of momentum carried across unit area of the plane in time dt by all molecules,

$$\tfrac{1}{3}\nu \left(\frac{\partial \overline{\mu}}{\partial z}\right) \int_0^\infty c\lambda_c f(c)\, dc\, dt = \tfrac{1}{3}\nu \left(\frac{\partial \overline{\mu}}{\partial z}\right) \overline{c\lambda_c}, \qquad \ldots\ldots(199)$$

where $\overline{c\lambda_c}$ denotes the mean value of $c\lambda_c$ averaged over all the molecules of the gas.

In the foregoing argument it might perhaps be thought that $\overline{\lambda_r}$ should have been replaced by $\tfrac{1}{2}\lambda_c$ instead of by λ_c. For if QO (fig. 31) is the whole free path described before collision occurs, there is no reason why PO should be less than PQ, so that the probable value of PQ might be thought to be $\tfrac{1}{2}\lambda$.

The fallacy in this reasoning becomes obvious on considering that after a molecule has left P, its chances of collision are exactly the same whether it has just undergone collision at P or has come undisturbed from Q. Hence $\overline{PO} = \lambda_c$, and therefore, by a similar argument, $\overline{PQ} = \lambda_c$.

A simple example taken from Boltzmann's *Vorlesungen*[*] will perhaps elucidate the point further. In a series of throws with a six-faced die the average interval between two throws of unity is of course five throws. But starting from any instant the average number of throws until a unit throw next occurs will be five, and similarly, working back from any instant, the average number of throws since a unit throw occurred is also five.

135. We may put

$$\overline{c\lambda_c} = \overline{c}l, \qquad \ldots\ldots(200)$$

where l is a new quantity, which is of course the mean free path of a molecule, this mean being taken in a certain way. This is not the same as any of the ways in which it was taken in the last chapter, so that we do not obtain an accurate result by replacing

[*] Vol. 1, p. 72.

l by any of the known values of the mean free path. At the same time the mean values calculated in different ways will not greatly differ from one another, and as our present calculation is at best one of approximation, we may be content for the moment to suppose *l* to be identical with the mean free path, however calculated. The extent of the error involved in this procedure will be examined later.

136. We have shewn that the aggregate transfer of momentum per unit of time across a unit area of a plane parallel to the plane of *xy* is

$$\tfrac{1}{3}v\bar{c}l\frac{\partial\bar{\mu}}{\partial z}. \qquad\qquad \ldots\ldots(201)$$

If we replace μ by its value mu, $\bar{\mu}$ becomes mu_0, and the transfer of momentum is seen to be

$$\tfrac{1}{3}\rho\bar{c}l\frac{\partial u_0}{\partial x}. \qquad\qquad \ldots\ldots(202)$$

This transfer of momentum results of course in a viscous drag of equal amount across the plane $z = z_0$. Now a viscous fluid, moving with the same velocity as the gas at every point, would exert a viscous drag

$$\eta\frac{\partial u_0}{\partial x}$$

per unit area across the plane $z = z_0$, where η is the coefficient of viscosity of the fluid. Thus the gas will behave exactly like a viscous fluid of viscosity η given by

$$\eta = \tfrac{1}{3}\rho\bar{c}l. \qquad\qquad \ldots\ldots(203)$$

137. We can now obtain some insight into the molecular mechanics of gaseous viscosity. Let us imagine two molecules, with velocities u, v, w and $-u, v, w$, penetrating from a layer at which the mass-velocity is 0, 0, 0 to one at which it is $u_0, 0, 0$. By the time the molecules have reached this second layer, we must suppose that their velocities are divided into two parts, namely,

$$u - u_0, v, w \quad \text{and} \quad u_0, 0, 0$$

for the first, and $\quad -u - u_0, v, w \quad$ and $\quad u_0, 0, 0$

for the second. The first part in each case will represent molecular-motion, and the second part will represent mass-motion. Now in § 92, it was found that the total energy of the gas could be regarded as the sum of the energies of the molecular and mass-motions; indeed, the sum of the energies of the molecular-motions of the two molecules now under discussion is easily seen to be

$$m(u^2 + v^2 + w^2) + mu_0^2.$$

The first term is equal to the energy of the molecular-motion of the two molecules at the start; the second term represents an increase which must be regarded as gained at the expense of the mass-motion of the gas. Thus we see that the phenomenon of viscosity in gases consists essentially in the degradation of the energy of mass-motion into energy of molecular-motion; this explains why it is accompanied by a rise of temperature in the gas.

Corrections when Molecules are assumed to be Elastic Spheres

138. From want of definite knowledge of molecular structure, two errors have been introduced into our calculations. In the first place we have neglected the persistence of velocities after collision, and in the second place we have ignored the difference between two different ways of estimating the mean free path. If the molecules are elastic spheres, it is possible to estimate the amount of error introduced by both these simplifications.

We may begin by an exact calculation of $\overline{c\lambda_c}$, to replace the assumption of equation (200). The quantity required is clearly

$$\overline{c\lambda_c} = \int_0^\infty f(c)\,\lambda_c c\,dc. \qquad \ldots\ldots(204)$$

On substituting the value for λ_c given by equation (164), and putting

$$f(c) = \sqrt{\frac{h^3 m^3}{\pi^3}}\,4\pi c^2 e^{-hmc^2},$$

we find

$$\overline{c\lambda_c} = \int_0^\infty \frac{4(hm)^{\frac{3}{2}} c^5 e^{-hmc^2}}{\pi v\sigma^2 \psi(c\sqrt{hm})}\,dc = \frac{4}{\pi\sqrt{hm}\,v\sigma^2}\int_0^\infty \frac{x^5 e^{-x^2}\,dx}{\psi(x)},$$

where $x = c\sqrt{hm} = \dfrac{1}{\sqrt{2}}\dfrac{c}{\bar{c}}$.

Thus if l is defined by equation (200), we must take

$$l = \frac{\overline{c\lambda_c}}{\bar{c}} = \overline{c\lambda_c}\sqrt{2hm} = \frac{8}{\sqrt{2}\pi\nu\sigma^2}\int_0^\infty \frac{x^5 e^{-x^2}dx}{\psi(x)}. \quad \ldots\ldots(205)$$

The integral can only be evaluated by quadratures. Tait* and Boltzmann† agree in assigning to l a value equal to 1·051 times Maxwell's mean free path calculated in § 108.

139. A more serious error has been introduced by neglecting the persistence of velocities which was investigated in the last chapter. When a molecule arrives at P after describing a path of which the projection on the axis of z is ζ, with a velocity of which the component parallel to the axis of z is w, then, on tracing back the motion, we know that as regards the previous path of each molecule the expectation of average velocity parallel to the axis of z is θw, where θ measures the persistence. The expectation of the projection of this path on the axis of z may therefore be taken to be $\theta\zeta$. Similarly, the expectation of the projection of each of the paths previous to these may be taken to be $\theta^2\zeta$, and so on. Thus the molecule must be supposed to have come, not from a distance ζ measured along the axis of z, but from a distance

$$\zeta + \theta\zeta + \theta^2\zeta + \ldots = \frac{\zeta}{1-\theta}. \quad \ldots\ldots(206)$$

We must not, however, assume that such a molecule on arriving at the plane $z = z_0$ has, on the average, a value of μ appropriate to the plane $z = z_0 + \dfrac{\zeta}{1-\theta}$. For the molecule has not travelled a distance $\dfrac{\zeta}{1-\theta}$ undisturbed, and at each collision a certain amount of its excess of momentum will have been shared with the colliding molecule. Of the various simple assumptions possible, the most obvious one to make is that at each collision the excess of momentum above that appropriate to the point at which the collision takes place is halved, half going to the colliding molecule and half remaining with the original molecule. Making this

---

*   *Collected Works*, **2**, pp. 152 and 178.
†   *Wien. Sitzungsber.* **84** (1881), p. 45.

assumption, it is clear that the excess of momentum to be expected is not that due to having travelled undisturbed a distance equal to that given by expression (206), but a distance

$$\zeta + \tfrac{1}{2}\{\theta\zeta + \tfrac{1}{2}[\theta^2\zeta + \tfrac{1}{2}(\theta^3\zeta + \ldots)]\} = \frac{\zeta}{1 - \tfrac{1}{2}\theta}. \qquad \ldots\ldots(207)$$

Thus we must insert a factor $1/(1 - \tfrac{1}{2}\theta)$ in the integrand of equation (205) before integration, the value of θ being obtained from the table on p. 151. As the result of a rough integration by quadratures, I find

$$l = \frac{1\cdot382}{\sqrt{2}\pi\nu\sigma^2}, \qquad \ldots\ldots(208)$$

so that the viscosity coefficient is given by

$$\eta = \tfrac{1}{3}\rho\bar{c}l = 0\cdot461\,\frac{m\bar{c}}{\sqrt{2}\pi\sigma^2}. \qquad \ldots\ldots(209)$$

140. This formula, although undoubtedly better than formula (203), is still only an approximation. It might be possible to improve still further on the rough assumptions just made, and so obtain results still closer to the truth. This, however, seems unnecessary, since exact numerical results are obtainable by the mathematical methods explained in Chap. IX below. There we shall see how Chapman, following Maxwell's method, has arrived at the exact formula

$$\eta = 0\cdot499\,\frac{m\bar{c}}{\sqrt{2}\pi\sigma^2}, \qquad \ldots\ldots(210)$$

and Enskog has confirmed this, by an entirely different method.

Variation of Viscosity with Density

141. Equation (210) shews that η is independent of the density of the gas, when the molecules are elastic spheres. And it is clear that whatever structure we assume for the molecules of the gas, l will, to a first approximation, vary inversely as the number of molecules per unit volume of the gas, so that formula (203) must give a value of η which is independent of ν. Increasing the number of molecules in a given volume increases the number of carriers, but decreases their efficiency as carriers *pari passu*; for on doubling the number of molecules per unit volume, the free path is reduced by half so that the momentum of any carrier will only differ, on the average, by half as much from the average momentum at the point of its next collision.

Thus we obtain Maxwell's law:

the coefficient of viscosity of a gas is independent of its density.

In spite of its apparent improbability, this law was predicted by Maxwell on purely theoretical grounds, and its subsequent experimental confirmation has constituted one of the most striking triumphs of the kinetic theory.

Some of the physical consequences of this law are interesting, and occasionally surprising. For instance, the well-known law of Stokes tells us that the final steady velocity v of a sphere falling through a viscous fluid is given by

$$v = \frac{g(M - M_0)}{6\pi a\eta},$$

where a, M are the radius and mass of the sphere, and M_0 the mass of fluid displaced. Since η is, by Maxwell's law, independent of the density, it follows that, within the limits in which Stokes's law is true, the final velocity of a sphere falling through air or any other gas will be independent of the density of the gas, or more strictly will depend on the density of the gas only through the term $M - M_0$, which will differ only inappreciably from M. Thus, a small sphere will fall as rapidly through a dense gas as through a rare gas. Again the air-resistances experienced by a pendulum ought to be independent of the density of the air, so that the oscillations of a pendulum ought to die away as rapidly in a rare gas as in a dense gas, as was in fact found to be the case by Boyle as far back as 1660.*

To test the truth of the general law, Maxwell† fixed three parallel and coaxal circular discs on a common axis, and then suspended them by a torsion thread in such a way that the three movable discs could oscillate between four parallel fixed discs.

---

* Thomson and Poynting, *Properties of Matter*, p. 218.
† *Phil. Trans.* **156** (1866), p. 249, or *Coll. Works*, 2, p. 1.

As in Boyle's pendulum experiment, the oscillations were found to die away at the same rate whether the air were dense or rare, up to pressures of one atmosphere.

142. At very much higher pressures at which the free path is becoming comparable only with molecular diameters, Maxwell's law fails altogether as is only to be expected. For we have supposed that a free path transports momentum through a distance λ with a velocity c, but we have so far overlooked that the collision which ends the free path transports it through a further distance σ with infinite velocity. Thus the transport for free path is through a distance of the form $\lambda + \sigma \cos \theta$.

Enskog* has shewn that, because of this, the value of η/ρ will not be constant, but will vary as

$$\frac{v}{b} + 0 \cdot 8000 + 0 \cdot 7614 \frac{b}{v},$$

where b and v mean the same as in Van der Waals' equation. This factor attains a minimum value of $2 \cdot 545$ when $v = 0 \cdot 8726b$, so that the general value of η/ρ can be put in the form.

$$\frac{\eta}{\rho} = \frac{1}{2 \cdot 545} \left[\frac{v}{b} + 0 \cdot 8000 + 0 \cdot 7614 \frac{b}{v} \right] \left(\frac{\eta}{\rho} \right)_{\min}.$$

This formula is found to represent observed variations of η/ρ fairly well, especially at high densities. The table on p. 166 gives the values of η for nitrogen at $50°$ C., as observed by Michels and Gibson,† together with the values calculated from Enskog's equation just given.

The minimum in η/ρ is clearly marked. The value $v = 0 \cdot 8726b$, at which theory predicts that it ought to occur, is reached at about 580 atmospheres.

Similar variations in η/ρ for carbon-dioxide have been observed by Warburg and v. Babo,‡ the minimum value being

---

* K. *Svenska Veten. Handb.* **63** (1922), No. 4; see also Chapman and Cowling, *The Mathematical Theory of Non-uniform Gases* (1939), p. 288.

† *Proc. Roy. Soc.* **134** A (1931), p. 307.

‡ *Wied. Ann.* **17** (1882), p. 390, and *Berlin. Sitzungsber.* (1882), p. 509. The numbers here given were subsequently corrected by Brillouin (*Leçons sur la viscosité des fluides,* 1907), and our text has reference to these corrected figures.

reached at a pressure of 77·2 atmospheres, at which the density is 0·450 gm. per cu. cm. Enskog has verified that these variations also are in accordance with his theoretical formula.

Viscosity of Nitrogen (50° C.) at high Pressures

Pressure (atmospheres)	η/ρ (observed)	$\eta \times 10^6$ (observed)	$\eta \times 10^6$ (calculated)
15·37	0·01179	191·3	181
57·60	0·003274	198·1	190
104·5	0·001928	208·8	205
212·4	0·001148	237·3	224
320·4	0·000952	273·7	266
430·2	0·000887	312·9	308
541·7	0·000866	350·9	348
630·4	0·000859	378·6	380
742·1	0·000870	416·3	418
854·1	0·000889	455·0	455
965·8	0·000909	491·3	492

A similar, although smaller, dependence of viscosity on density has also been observed in hydrogen at moderate pressures by Kamerlingh Onnes, Dorsman and Weber.*

At the other limit of excessively small pressure, remarkable departures from Maxwell's law may occur through the free path becoming comparable with, or even greater than, the dimensions of the vessel in which the experiment is conducted. If the molecule has not room to describe a free path equal to the theoretical free path assumed in § 140, the resulting formula obtained for the viscosity must obviously fail. If l cannot, from the arrangement of the apparatus, be greater than some value l_0, then η (cf. equation (209)) cannot be greater than $\frac{1}{3}\rho\bar{c}l_0$, and so ought to vanish with ρ. This is found to be the case. As far back as 1881, Sir W. Crookes† measured the viscosities of gases at pressures of only a few thousandths of a millimetre of mercury, and obtained values much smaller than those at higher pressures, which tended to vanish altogether as the density of the gas vanished. Later investigators have abundantly confirmed his conclusions.

* Comm. Phys. Lab. Leiden, 134a (1913). See also Winkelmann's Handbuch der Physik (IIte aufl.), pp. 1399, 1406.

† Phil. Trans. 172 (1881), p. 387.

A case of special interest, which we shall now discuss, occurs when gas flows through a tube of small or capillary cross-section.

The Flow of Gas through Tubes

143. When gas flows through a tube, the layer of gas in contact with the wall of the tube is usually held at rest by the tube, while the rate of flow increases as we pass inwards towards the centre of the tube.

Let us suppose the tube to be of circular cross-section, of radius R, and length L. Let the gas at any distance r from the axis be flowing along the tube with a velocity v.

The cylinder of gas of radius r coaxal with the tube will experience a viscous drag $-\eta \dfrac{dv}{dr}$ per unit area over its whole surface, and as this is of area $2\pi r L$, the whole cylinder experiences a viscous drag of amount

$$-2\pi r L \eta \frac{dv}{dr}.$$

If the flow is steady, so that the gas moves without acceleration, this viscous drag must be exactly balanced by the difference of the pressures at the two ends of the cylinder. This will be $\pi r^2 (p_1 - p_2)$, where p_1, p_2 are the pressures per unit area at the ends of the tube, so that $p_1 - p_2$ is the pressure-difference employed to drive the gas through the tube. Thus

$$-2\pi r L \eta \frac{dv}{dr} = \pi r^2 (p_1 - p_2).$$

Dividing throughout by $-2\pi r L \eta$ and integrating, we obtain

$$v = A - \frac{p_1 - p_2}{4L\eta} r^2, \qquad \ldots\ldots(211)$$

where A is a constant of integration.

If there is no slip between the gas and the walls of the tube, v must be zero when $r = R$. Determining A from this condition, equation (211) becomes

$$v = \frac{p_1 - p_2}{4L\eta} (R^2 - r^2). \qquad \ldots\ldots(212)$$

The total volume V of gas which flows through the tube in unit time is

$$V = \int_0^R 2\pi r v \, dr = \frac{\pi(p_1 - p_2)}{8L\eta} R^4. \qquad \ldots\ldots(213)$$

This is Poiseuille's formula for the flow of gas through a tube. Since the quantities V, L, R and $p_1 - p_2$ all admit of easy measurement, it provides a convenient method for measuring η experimentally.

Experiment shews, however, that the flow of gas in tubes of very small diameter is often greater than that given by this formula, as though the gas slipped at its contact with the walls of the tube. In 1860 Helmholtz and Piotrowski* shewed that a slip of this kind actually occurred in liquids, and later Kundt and Warburg† shewed the same for gases. A great number of investigators, starting with Maxwell,‡ have discussed the theory of this slip.

If such a slip occurs, let v_0 be the velocity of the layer of gas which is in contact with the walls of the tube. This slip will produce a viscous force on the gas which will be jointly proportional to v_0 and to the area of surface, namely $2\pi RL$, over which slip occurs. We may take the whole viscous drag on the gas to be $2\pi RL\epsilon v_0$, where ϵ is a constant. Equating this to the force urging the whole gas through the tube,

$$2\pi RL\epsilon v_0 = \pi R^2 (p_1 - p_2),$$

so that
$$v_0 = \frac{p_1 - p_2}{2L\epsilon} R.$$

The constant A in equation (211) must now be adjusted so as to give this value for v when $r = R$. We accordingly find, as the general value of v,

$$v = \frac{p_1 - p_2}{4L\eta}\left(R^2 - r^2 + \frac{2R\eta}{\epsilon}\right),$$

and, instead of equation (213), for the flow per unit time,

$$V = \frac{\pi(p_1 - p_2)}{8L\eta} R^4\left(1 + \frac{4\eta}{\epsilon R}\right). \qquad \ldots\ldots(214)$$

---

* *Wien. Sitzungsber.* **40** (1860), p. 607.

† *Pogg. Annalen,* **155** (1875), p. 337.   ‡ *Coll. Works,* **2**, p. 703.

We see that the slip at the walls increases the flow by a fraction $4\eta/\epsilon R$. This is generally known as the Kundt and Warburg correction. It is unimportant so long as R is large in comparison with η/ϵ, but when R is small compared with η/ϵ, it takes control of the whole process, and results in the flow being proportionate to R^3 instead of to R^4.

144. The quantity η/ϵ, which measures the ratio of the friction of the gas on itself to the friction between the gas and the solid, is generally called the "coefficient of slip". Maxwell* gave much thought to the problem of its evaluation. He assumed in the first instance that of the molecules which impinged on a solid wall only a fraction $1-f$ are reflected back "specularly", i.e. at the same angle as that at which they struck the wall, the remaining fraction f leaving the wall at random angles which are independent of the earlier motions of the molecule.

From a discussion of the experiments of Kundt and Warburg, Maxwell concluded that f must be about $\frac{1}{2}$. The experiments of Knudsen described in § 34 seem, on the contrary, to indicate that, under some circumstances at least, the correct value of f is not far from unity; most or all of the molecules start out in purely random directions after impact.

Blankenstein† has directly measured the coefficients of slip between a number of gases and polished oxydized silver, and finds for f the values 1·00, 1·00, 0·99 and 0·98 for helium, hydrogen, oxygen and air respectively. For reflection from other solids the value of f may be substantially smaller. Thus Millikan‡ gives the values

Air or CO_2 on machined brass, old shellac, or mercury $f = 1\cdot00$

Air on oil   0·895

CO_2 on oil   0·92

Air on glass   0·89

Air on fresh shellac   0·79

145. The Kundt and Warburg formula (214) is found to agree well with experiment so long as the pressure of the gas is sufficiently high, but fails entirely, as might be anticipated,

* *Coll. Works*, 2, p. 703. See also Loeb, *Kinetic Theory of Gases* (2nd ed., 1934), pp. 285 ff.

† *Phys. Rev.* 22 (1923), p. 582.   ‡ *Phys. Rev.* 21 (1923), p. 217.

for pressures so low that the free path is comparable with, or greater than, the diameter of the tube. In this case it is remarkable that the rate of flow is independent of both the viscosity and the density of the gas. The appropriate analysis has been given by Knudsen.*

We may suppose that the velocity with which the gas flows through the tube is small in comparison with the molecular velocities, and that these latter conform to Maxwell's law for a gas with a mass-velocity equal to the velocity of flow. The total mass of the molecules which impinge on unit area of the wall of the tube in unit time is thus $\frac{1}{4}\rho\bar{c}$, as in § 39. The average velocity along the tube is now the same at all points of the gas, say u_0. Thus the total momentum that the moving molecules give up to the walls of the tube in unit time is $2\pi RL \times \frac{1}{4}\rho\bar{c}u_0$. This, as before, must be equal to $\pi R^2(p_1-p_2)$, so that

$$u_0 = \frac{2R(p_1-p_2)}{L\rho\bar{c}}.$$

The total mass of gas which flows through the tube in unit time is accordingly†

$$\pi R^2 \rho u_0 = \frac{2R^3(p_1-p_2)}{L\bar{c}}, \qquad \ldots\ldots(215)$$

which, as already remarked, is independent of both η and ρ. Knudsen has verified that this formula agrees well with observation. Comparing it with Poiseuille's formula, we notice that the flow increases as the cube instead of as the fourth power of R.

VARIATION OF VISCOSITY WITH TEMPERATURE

146. We turn next to examine how the viscosity of a gas varies with its temperature.

Since $\bar{c}$ is proportional to the square root of the absolute temperature, formula (210) shews that if the molecules were true

---

* *Ann. d. Phys.* **28** (1909), pp. 75, 999; **32** (1910), p. 809; **34** (1911), p. 593, and later papers. See also M. Knudsen, *The Kinetic Theory of Gases* (Methuen, 1934), pp. 21 ff.

† Knudsen gives a formula which is less than this (erroneously as it seems to me) by a factor $\dfrac{8}{3\pi}$, but the experimental tests agree with this formula better than with formula (215).

elastic spheres, the value of η would be proportional to the square root of the temperature.

As a matter of fact, η is found to vary a good deal more rapidly than this as the temperature increases. The divergence between experiment and the theoretical value obtained on the assumption that the molecules are elastic spheres is, however, one that could have been predicted. The assumption in question is, at best, only an approximation, and we must continually examine what deviations are to be expected from the results to which it leads.

The peculiarity of a system of elastic spheres is that the motion remains geometrically the same if the velocity of every sphere is increased in the same ratio. If the molecules are surrounded by repulsive fields of force, this is no longer the case; increasing the velocities increases the degree to which the molecules penetrate into each others fields of force at collision, and so has the result of decreasing the effective sizes of the molecules.

Thus if the molecules of a gas are surrounded by fields of force, and we attempt to represent these molecules by elastic spheres, we must suppose the size of these spheres to vary with the temperature of the gas. The spheres are large at low temperatures, small at high.

It follows that in formula (210), η must be supposed to depend on the temperature both through the factor $\bar{c}$ in the numerator, and also through the factor σ^2 in the denominator. Thus the value of η will not vary simply as the square root of the temperature, but will vary more rapidly with the temperature than this.

From the way in which η is observed to vary with the temperature, we can obtain some information as to the fields of force surrounding the molecules. For from equation (210) we can calculate σ as a function of the temperature T. Now the value of σ at any specified temperature is, roughly speaking, the average distance of closest approach of the centres of two molecules in collision, so that the mutual potential energy of two molecules at a distance σ is, on the average, equal to the kinetic energy of the velocities along the line of centres before collision.

We readily find (formula (140)) that the average value of V^2, the square of the relative velocity before collision, is $\frac{8}{3}C^2$. Thus the square of the velocity of each molecule relatively to the centre of gravity of the two colliding molecules will be, on the average, $\frac{2}{3}C^2$. The probability that this velocity is at an angle between θ and $\theta + d\theta$ with the line of centres is $2\sin\theta\cos\theta\,d\theta$, so that the average square of the relative velocity along the line of centres $V\cos\theta$ is

$$\frac{2}{3}C^2 \int_0^{\frac{1}{2}\pi} 2\sin\theta\cos^3\theta\,d\theta = \frac{1}{3}C^2.$$

The kinetic energy which has been destroyed by the inter-molecular field of force when the molecules are, on the average, at their point of closest approach at distance σ apart is therefore $\frac{1}{3}mC^2$ or RT. Thus the mutual potential energy of two molecules at a distance σ apart will be RT, where T is the temperature corresponding to the value of σ in question. The force of repulsion between two molecules at a distance σ is accordingly $-R\dfrac{dT}{d\sigma}$.

If the law of force is μr^{-s}, we must have

$$\frac{\mu}{\sigma^s} = -R\frac{dT}{d\sigma},$$

giving on integration

$$\sigma = \left[\frac{\mu}{RT(s-1)}\right]^{\frac{1}{s-1}}. \qquad \ldots\ldots(216)$$

Here σ denotes the distance of closest approach of two molecules at an encounter, and when the orbits are at all curved, this is not quite the same thing as the diameter of the sphere obtained by supposing the molecules to be elastic spheres. Thus equation (216) will give a value of σ which will differ by a numerical multiplier from the value obtained in § 54 by considering the deviations from Boyle's law. This multiplier will of course vary for different values of s. It will reduce to unity for elastic spheres, and will differ most from this for the smallest values of s.

In § 54 we found that molecules with a law of force μr^{-s} could be regarded as elastic spheres for the purpose of calculating the pressure, if σ were supposed given by

$$\sigma = \left[\frac{\mu}{RT(s-1)}\right]^{\frac{1}{s-1}} \sqrt[3]{\Gamma\left(1 - \frac{3}{s-1}\right)}. \qquad \ldots\ldots(217)$$

This agrees with (216) except for the numerical factor, and agrees completely, as it ought, for elastic spheres ($s = \infty$). We see that molecules which are point centres of force may be treated as elastic spheres, both as regards pressure and viscosity, but the spheres must be of different sizes in the two cases, except of course when the molecules really are spheres.

Although formulae (216) and (217) become identical when s is infinite, the divergence between them may be very considerable when s is small. The lowest value for s which can be supposed to occur for any gas is probably about $s = 5$ (cf. § 147, below), and when $s = 5$,

$$\sqrt[3]{\Gamma\left(1 - \frac{3}{s-1}\right)} = \sqrt[3]{\Gamma\left(\frac{1}{4}\right)} = 1{\cdot}5363.$$

Thus for such a gas as carbon-dioxide, for which $s = 5{\cdot}6$, we may expect a difference of as much as 50 per cent between the values of σ calculated from viscosity and Boyle's law.

In such a case as this, however, the calculation from Boyle's law fails because b, which from equation (212) ought to vary as $T^{-\frac{2}{3}}$, is supposed, in evaluating b experimentally, to remain independent of the temperature.

Whatever the value of the numerical multiplier may be, it appears that $\dfrac{1}{\sigma^2}$ will vary as $T^{\frac{2}{s-1}}$, so that η will vary as T^n, where

$$n = \tfrac{1}{2} + \frac{2}{s-1}. \qquad\qquad \ldots\ldots(218)$$

147. For some gases, η is found as a matter of experiment to vary approximately as a power of T, being represented with very tolerable accuracy by the formula

$$\eta = \eta_0 \left(\frac{T}{273{\cdot}2}\right)^n, \qquad\qquad \ldots\ldots(219)$$

where η_0 is of course the coefficient of viscosity at $0°$ C. A good instance is helium; the agreement between the observed values of η for this gas and the values given by formula (215) is exhibited in the table on p. 179 below.

Clearly the molecules of substances for which there is a good agreement of this kind may be regarded as point centres of force, repelling according to the law μ/r^s, where s is given by equation (218).

If a gas is constituted of molecules which are not of this type, it will still be possible to represent the viscosity-coefficient by a formula of the type of (219), through any small range of temperature we please. For the value of n is at our disposal, and may be chosen so as to give the right value to $d\eta/dT$, the slope of the viscosity-coefficient, through this small range. But this value of n will not usually satisfy the experimental data through another range; for this some other value of n must be chosen.

The following table gives the values of n for a number of gases through ranges which include 0° C., together with the

Values of n and s for Certain Gases

Gas	Authority*	Value of n (observed)	Value of s (calculated)
Hydrogen	1	0·695	11·3
Deuterium	2	0·699	11·0
Helium†	3	0·647	14·6
Neon	4	0·657	13·7
Nitrogen	5	0·756	8·8
Carbon-monoxide	6	0·758	8·75
Air	6	0·768	8·46
Oxygen	5	0·814	7·40
Hydrochloric Acid	4	1·03	4·97
Argon	7	0·823	7·19
Nitrous oxide	6	0·89	6·15
Carbon-dioxide	6	0·935	5·6
Chlorine	8	1·0	5·0

* Authorities:
1. Kamerlingh Onnes, Dorsman and Weber, *Verslag. Kon. Akad. van Wetenschappen*, Amsterdam, **21** (1913), p. 1375.
2. Van Cleave and Maas, *Canadian Journ. of Research*, **13**B (1935), p. 384.
3. Kamerlingh Onnes and Weber, *Verslag. Kon. Akad. van Wetenschappen*, Amsterdam, **21** (1913), p. 1385.
4. Trautz and Binkele, *Ann. d. Phys.* **5** (1930), p. 561.
5. Markowski, *Ann. d. Phys.* **14** (1904), p. 742.
6. Values given by Chapman and Cowling (*The Mathematical Theory of Non-uniform Gases*, Cambridge, 1939), calculated from data of various experimenters.
7. Schultze, *Ann. d. Phys.* **5** (1901), p. 163 and **6** (1901), p. 301.
8. Trautz and Winterkorn, *Ann. d. Phys.* **10** (1931), p. 522.
† See § 148, below.

values of s calculated from relation (218). An instance of the closeness of agreement between formula (219) and observation will be found below (§ 149).

If the molecules of a gas were in actual fact elastic spheres, the value of n would of course be zero, and s would be infinite. Molecules for which s is large are frequently described as "hard", and those with smaller values of s as "soft". Our table shews that, generally speaking, the hardest molecules are those of simplest structure. The monatomic molecules helium and neon come first, then diatomic molecules of simple structure— hydrogen and deuterium, followed by nitrogen, carbon-monoxide and oxygen. The softest molecules are those of chlorine and hydrochloric acid for which s is nearly five. This value seems to constitute a sort of natural lower limit for s.

148. We next examine the absolute value of η. We have already seen that η is proportional to $m\bar{c}/\sigma^2$, or to $\sqrt{mRT}/\sigma^2$. Using the value of σ given by equation (216), we find that the coefficient of viscosity must be given by an equation of the form

$$\eta = A \sqrt{mRT}\left[\frac{RT(s-1)}{\mu}\right]^{\frac{2}{s-1}}, \qquad \ldots\ldots(220)$$

where A is a numerical constant.

Chapman* has determined the value of this constant by very complicated analysis. To a first approximation he found its value to be

$$A = \frac{5\sqrt{\pi}}{8I_2(s)\,\Gamma\left(4-\dfrac{2}{s-1}\right)}, \qquad \ldots\ldots(221)$$

where $I_2(s)$ is a pure number depending only on s, its actual value being

$$I_2(s) = \pi\int_0^\infty \sin^2\theta'd\,d\alpha,$$

where θ' and α have the meanings assigned to them in § 196 below. This integral was introduced by Maxwell† in 1866. He calculated that when $s = 5$ its value is $I_2(5) = 1\cdot3682$, and shewed that when $s = 5$ formula (221) is exact.

* *Phil. Trans. Roy. Soc.* **211 A** (1912), p. 433.
† *Coll. Works,* **2**, p. 42.

In a later paper*, Chapman carries the calculations to a second approximation, and finds that the value of A given by equation (221) must be multiplied by a factor which increases continuously from unity when $s = 5$ to $1 \cdot 01485$ when $s = \infty$. Thus the error in using approximation (221) for A is never more than about $1\frac{1}{2}$ per cent, and as this is smaller than experimental errors of observation, it is hardly worth carrying the approximation further. Before leaving this question, it may be remarked that Enskog† has given the factor by which A must be multiplied when the molecules repel as the inverse sth power of the distance in the form

$$1 + \frac{3(s-5)^2}{2(s-1)(101s-113)} + \dots .$$

This of course reduces to unity when $s = 5$ and has the value $1\frac{3}{202}$ or $1 \cdot 01485$ when $s = \infty$, thus agreeing with Chapman's value just quoted.

For hard spherical molecules ($s = \infty$) these formulae lead to the value of η already given in formula (210).

Sutherland's molecular model

149. The foregoing theory has been concerned only with molecules which attract or repel according to the law μr^{-s}. It cannot be supposed that any actual molecules are of so simple a type as this.

For, as we have already seen, the supposition that the molecular force falls off as an inverse power of the distance leads to formula (216) which requires σ to vanish absolutely at very high temperatures. It seems more natural, and more in accordance with modern knowledge of molecular structure, to suppose that a molecule possesses a hard kernel which is not penetrated by other molecules no matter how violent the collision between them may be.

---

* *Phil. Trans. Roy. Soc.* **216** A (1915), p. 279.

† *Kinetische Theorie der Vorgänge in mässig verdünnten Gasen* (Inaug. Dissertation, Upsala, 1917). Enskog's analysis is given by Chapman and Cowling, in a somewhat different form, in their book *The Mathematical Theory of Non-uniform Gases* (C.U.P., 1939).

As far back as 1893, Sutherland* had assumed that the effective value of σ at temperature T is

$$\sigma^2 = \sigma_\infty^2\left(1 + \frac{C}{T}\right),$$

where C, σ_∞ are constants, σ_∞ being the value of σ when $T = \infty$, and therefore being the diameter of the hard kernel of the molecule, while C is the temperature at which $\sigma^2 = 2\sigma_0^2$.

As the temperature varies, the coefficient of viscosity varies as $\bar{c}l$, and so as $\dfrac{T^{\frac{1}{2}}}{\sigma^2}$, or again as $\dfrac{T^{\frac{3}{2}}}{C+T}$. Thus if η_0 is the coefficient of viscosity at $0°$ C. ($T = 273\cdot2$), the general coefficient of viscosity η at temperature T will be given by

$$\eta = \eta_0\left(\frac{T}{273\cdot2}\right)^{\frac{3}{2}}\frac{C+273\cdot2}{C+T},$$

which is Sutherland's formula for the viscosity at temperature T.

For many gases this formula meets with very considerable success in predicting the variation of viscosity with temperature. As an illustration may be given the following tables, taken from a paper by Breitenbach,† in which the observed and calculated values of the viscosity are compared.

Ethylene
($\eta_0 = 0\cdot00009613$, $C = 225\cdot9$)

Temperature	η (observed)	η (calculated)
$-21\cdot2°$ C.	0·0000891	0·0000890
15·0	1006	1012
99·3	1278	1278
182·4	1530	1519
302·0	1826	1833

Carbon-dioxide
($\eta_0 = 0\cdot00013879$, $C = 239\cdot7$)

Temperature	η (observed)	η (calculated)
$-20\cdot7°$ C.	0·0001294	0·0001284
15·0	1457	1462
99·1	1861	1857
182·4	2221	2216
302·0	2682	2686

* *Phil. Mag.* **36** (1893), p. 507.   † *Ann. d. Physik,* **6** (1901), p. 168.

The following values for C have been found by different observers:

Helium	$C = 80·3$ (Schultze), $78·2$ (Schmitt), 70 (Rankine).
Neon	$C = 56$ (Rankine).
Argon	$C = 169·9$ (Schultze), $174·6$ (Schmitt), 142 (Rankine).
Krypton	$C = 188$ (Rankine).
Xenon	$C = 252$ (Rankine).
Hydrogen	$C = 72·2$ (Rayleigh), $71·7$ (Breitenbach), 79 (Sutherland), 83 (Schmitt).
Nitrogen	$C = 102·7$ (Trautz and Baumann), 118 (Smith).
Carbon-monoxide	$C = 100$ (Sutherland), 118 (Smith).
Air	$C = 111·3$ (Rayleigh), $119·4$ (Breitenbach), 113 (Sutherland).
Nitric oxide	$C = 128$ (Trautz and Gabriel).
Oxygen	$C = 138$ (Markowski), 138 (Schmitt).
Chlorine	$C = 325$ (Rankine).
Nitrous oxide	$C = 260$ (Sutherland), 274 (Smith).
Carbon-dioxide	$C = 239·7$ (Breitenbach), 277 (Sutherland), 274 (Smith).
Ethylene	$C = 225·9$ (Breitenbach), 272 (Sutherland).
Methyl chloride	$C = 454$ (Breitenbach).

Authorities:

Schultze, *Ann. d. Phys.* **5** (1901), p. 165, and **6** (1901), p. 310.
Schmitt, *Ann. d. Phys.* **30** (1909), p. 398.
Rankine, *Proc. Roy. Soc.* **84** A (1910), p. 188 and **86** A (1912), p. 162.
Rayleigh, *Proc. Roy. Soc.* **66** A (1899), p. 68 and **67** A (1900), p. 137.
Breitenbach, *Ann. d. Phys.* **5** (1901), p. 168.
Sutherland, *Phil. Mag.* **36** (1893), p. 507.
Smith, *Proc. Phys. Soc.* **34** (1922), p. 155.
Trautz and Baumann, *Ann. d. Phys.* **2** (1929), p. 733.
Trautz and Gabriel, *Ann. d. Phys.* **11** (1931), p. 607.
Markowski, *Ann. d. Phys.* **14** (1904), p. 742.

We should obviously expect the more permanent gases, which have low critical temperatures and low boiling points, to have low values for the Sutherland temperature C. Rankine[*] has noticed that most gases for which data are available have critical temperatures equal to about $1·14$ times C.

On the other hand, Kamerlingh Onnes[†] finds very definitely that the viscosity of helium at low temperatures cannot be represented by Sutherland's formula with anything like the

[*] *Proc. Roy. Soc.* **86** A (1912), p. 166.
[†] Kamerlingh Onnes and Sophus Weber, *Comm. Phys. Lab. Leiden*, **134** b, p. 18.

accuracy given by the simpler formula (219). This is shewn in the following table: the second column gives the values of η observed for helium, the third column gives the values calculated from formula (219) on taking $\eta_0 = 0.0001887$, $n = 0.647$, while the last column gives values of η calculated by Sutherland's formula, taking $C = 78.2$.

Viscosity of Helium

Temperature	η (observed)	$\eta_0\left(\dfrac{T}{273.2}\right)^{0.647}$	η (calculated, Sutherland)
183·7° C.	0·0002681	0·0002632	0·0002682
99·8	2337	2309	2345
18·7	1980	1970	1979
17·6	1967	1965	1974
− 22·8	1788	1783	1771
− 60·9	1587	1603	1563
− 70·0	1564	1558	1513
− 78·5	1506	1515*	1460
− 102·6	1392	1389	1317
− 183·3	09186	09185	0745
− 197·6	08176	08213	0628
− 198·4	08132	08155	0621
− 253·0	03498	03489	0135
− 258·1	02946	02887	0092

* This entry, which was obviously wrong in the original table, has been recalculated.

Very similar results have also been obtained for hydrogen by Kamerlingh Onnes, Dorsman and Weber,[†] while a general failure of Sutherland's formula to represent viscosity at low temperatures has been noticed and discussed by Schmitt, Bestelmeyer, Vogel and others.[‡]

More General Laws of Force

150. This has led to a study of the way in which viscosity would depend on temperature with more general laws for the forces between molecules. In 1924 J. E. Lennard Jones[§] introduced a tentative law of force

$$f(r) = \frac{\lambda_n}{r^n} - \frac{\lambda_3}{r^3}$$

[†] *Comm. Phys. Lab. Leiden*, **134a** (1913).
[‡] For references see Chapman, *Phil. Trans.* **216 A** (1915), p. 342.
[§] *Proc. Roy. Soc.* **106 A** (1924), p. 441, and subsequent papers.

and examined the resulting dependence of viscosity on temperature. This was soon discarded in favour of the more general law

$$f(r) = \frac{\lambda}{r^n} - \frac{\mu}{r^m}.$$

Hassé and Cook* have calculated the resulting formula for the
coefficient of viscosity in the special case of $n = 9$, $m = 5$. They
found that, with a suitable choice of values for λ and μ, the
formula could be made to fit the experiments well for hydrogen,
nitrogen and argon, but could not be made to fit for helium,
carbon-dioxide and neon. The agreement of their formula with
observation for hydrogen and argon is shewn in the two tables
below.

Viscosity of Hydrogen

Abs. temp.	$\eta \times 10^7$ (observed)	$\eta \times 10^7$ (Hassé and Cook)	$\eta \times 10^7$ (Kamerlingh Onnes)
457·3	1212	1226	1207
373·6	1046	1060	1052
287·6	877	878	875
273·0	844	846	843
261·2	821	820	816
255·3	802	806	803
233·2	760	756	757
212·9	710	708	709
194·4	670	664	666
170·2	609·3	603	608
89·63	392·2	380	389
70·87	319·3	320	329
20·04	105–111	111	137

Here the second column gives the best observed value of the
viscosity coefficient, while the third column gives the value
calculated by Hassé and Cook. The last column gives the value
calculated by Kamerlingh Onnes for a simple repulsive force
varying as $r^{-11 \cdot 2}$, and it will be seen that the agreement with
observation is not enormously less good than with the more
complex law of Hassé and Cook.

* H. R. Hassé and W. R. Cook, *Proc. Roy. Soc.* **125** A (1929), p. 196.

Viscosity of Argon

Abs. temp.	$\eta \times 10^7$ (observed)	$\eta \times 10^7$ (Hassé and Cook)
456·4	3243	3212
372·8	2751	2745
286·3	2207	2216
272·9	2116	2129
252·8	1987	1993
232·9	1854	1855
212·9	1697	1711
194·3	1575	1572
168·7	1379	1373
89·9	735·6	686·9

VISCOSITY IN A MIXTURE OF GASES

151. As the proportions of two kinds of gas in a mixture change from $1:0$ to $0:1$, the coefficient of viscosity of the mixture will of course also change, starting from the coefficient of viscosity η_1 of the first gas, and ending at the coefficient of viscosity of the second gas η_2. But, as Graham* found in 1846, the change may not be continuous, and for certain proportions of the mixture the coefficient of viscosity of the mixture η_{12} may have a value greater than either of the coefficients of viscosity η_1, η_2 of the pure gases.

To a first rough approximation, the viscous transfer of momentum in a mixture of gases may be regarded as the sum of the transfers by the different kinds of molecules separately. Thus if $\rho_1, \rho_2, \rho_3, \ldots$ are the densities of different kinds of gas, $\bar{c}_1, \bar{c}_2, \bar{c}_3, \ldots$ the average velocities of the molecules of these different kinds and $\lambda_1, \lambda_2, \lambda_3, \ldots$ the average free paths, we may expect the viscosity of the mixture to be given by

$$\eta = \tfrac{1}{3}(\rho_1\bar{c}_1\lambda_1 + \rho_2\bar{c}_2\lambda_2 + \rho_3\bar{c}_3\lambda_3 + \ldots).$$

For a simple binary mixture, we should expect a coefficient of viscosity η_{12} given by

$$\eta_{12} = \tfrac{1}{3}\rho_1\bar{c}_1\lambda_1 + \tfrac{1}{3}\rho_2\bar{c}_2\lambda_2,$$

or, substituting for λ_1 and λ_2 from formula (151),

$$\eta_{12} = \frac{\eta_1}{1 + A_1\dfrac{\rho_2}{\rho_1}} + \frac{\eta_2}{1 + A_2\dfrac{\rho_1}{\rho_2}}, \qquad \ldots\ldots(222)$$

* *Phil. Trans. Roy. Soc.* **136** (1846), p. 573.

where η_1, η_2 are the coefficients of viscosity of the two constituents when pure, and A_1, A_2 are quantities which depend on the masses and diameters of the molecules of the two components.

When ρ_2/ρ_1 is zero, the value of η_{12} given by this formula reduces of course to η_1; when ρ_2/ρ_1 is small, it is

$$\eta_{12} = \eta_1 + \frac{\rho_2}{\rho_1}\left(\frac{\eta_2}{A_2} - \eta_1 A_1\right).$$

Thus if $\eta_2 > \eta_1 A_1 A_2$, a small admixture of the second gas increases the coefficient of viscosity. If this inequality is satisfied and η_2 is also less than η_1, the value for η_{12} must clearly rise to a maximum as ρ_2/ρ_1 is increased and then descend again to its final value η_2.

A formula of this type was first given by Thiesen.* Schmitt,† and later Schroer,‡ have found that such a formula represents many, or even most, of the experimental data on the viscosity of binary mixtures with very fair accuracy.

The exact theoretical investigation of viscosity in a mixture of gases is very complicated. Formulae for the coefficients of viscosity of mixture have been given by Maxwell, Kuenen, Chapman and Enskog, and in every case the theory predicts a maximum value for a certain ratio of the gases in accordance with observation. Maxwell's investigation§ deals only with molecules repelling as the inverse fifth power of the distance; Kuenen‖ deals with elastic spheres, the formulae being corrected for the phenomenon of "persistence of velocity" explained in Chap. v; while Chapman¶ and Enskog** discuss the viscosity of gas-mixture by following the general methods which are explained in Chap. ix below.

Chapman's analysis shews that a formula of the type of (222) must necessarily fail to represent the facts with any completeness;

---

* *Verhand. d. Deutsch. Phys. Gesell.* **4** (1902), p. 238.

† *Ann. d. Physik*, **30** (1909), p. 303.

‡ *Zeitschr. f. Phys. Chem.* **34** (1936), p. 161.　　§ *Coll. Papers*, **2**, p. 72.

‖ *Proc. Konink. Akad. Wetenschappen, Amsterdam*, **16** (1914), p. 1162 and **17** (1915), p. 1068.

¶ *Phil. Trans.* **216** A (1915), p. 279, and **217** A (1916), p. 115; *Proc. Roy. Soc.* **93** A (1916), p. 1.

** *Kinetische Theorie der Vorgänge in mässig verdünnten Gasen*, Inaug. Dissertation, Upsala, 1917.

the necessary formula is found to be of the more general type

$$\eta_{12} = \frac{\eta_1 a_1 \rho_1^2 + a_{12}\rho_1\rho_2 + \eta_2 a_2\rho_2^2}{a_1\rho_1^2 + b\rho_1\rho_2 + a_2\rho_2^2},$$

where a_1, a_{12}, a_2 and b are new constants depending on the molecular masses, the law of force, and the temperature. Chapman shewed that a formula of this type represented the observations of Schmitt,[†] and later Trautz[‡] and his collaborators have established further agreement between this formula and experiment.

Gas	Mol. wt.	η (observed) at 0° C.	Authority*	$\frac{1}{2}\sigma$ (calc.) in cm.
Monatomic				
Helium	4	0·000188	1	$1·09 \times 10^{-8}$
Neon	20	0·000312	1	$1·30 \times 10^{-8}$
Argon	40	0·000210	1	$1·83 \times 10^{-8}$
Krypton	84	0·000233	1	$2·08 \times 10^{-8}$
Xenon	131	0·000211	1	$2·46 \times 10^{-8}$
Mercury	201	0·000162	2	$3·13 \times 10^{-8}$
Diatomic				
Hydrogen	2	0·00008574	3	$1·36 \times 10^{-8}$
Carbon-monoxide	28	0·0001665	4	$1·89 \times 10^{-8}$
Nitrogen	28	0·000167	2	$1·89 \times 10^{-8}$
Air	—	0·000172	2	$1·87 \times 10^{-8}$
Nitric oxide	30	0·0001794	5	$1·88 \times 10^{-8}$
Oxygen	32	0·000192	2	$1·81 \times 10^{-8}$
Hydrogen sulphide	33	0·000118	6	$2·32 \times 10^{-8}$
Chlorine	71	0·000122	7	$2·70 \times 10^{-8}$
Polyatomic				
Water vapour	18	0·000087	2	$2·33 \times 10^{-8}$
Nitrous oxide	44	0·0001366	4	$2·33 \times 10^{-8}$
Carbon-dioxide	44	0·000137	2	$2·33 \times 10^{-8}$
Methane (CH_4)	16	0·000108 (20° C.)	2	$2·04 \times 10^{-8}$
Benzene (C_6H_6)	78	0·0000700	2	$3·75 \times 10^{-8}$

* Authorities:
1. Rankine, *Proc. Roy. Soc.* 83 A (1910), p. 516; 84 A (1910), p. 181.
2. Kaye and Laby, *Physical Constants* (8th edn., 1936).
3. Breitenbach, *Ann. d. Phys.* 5 (1901), p. 166.
4. C. J. Smith, *Proc. Phys. Soc.* 34 (1922), p. 155.
5. Eucken, *Phys. Zeitschr.* 14 (1913), p. 324.
6. Rankine and Smith, *Phil. Mag.* 42 (1921), pp. 601, 615.
7. Rankine, *Proc. Roy. Soc.* 86 A (1912), p. 162.

† *L.c. ante.*
‡ *Ann. d. Physik*, 3 (1929), p. 409; 7 (1930), p. 409; 11 (1931), p. 606.

DETERMINATION OF SIZE OF MOLECULES

152. We have already, in § 31, had an instance of the calculation of molecular radii from the coefficients of viscosity. When the coefficient of viscosity of any gas has been determined by experiment, it is possible to regard equation (210) as an equation for σ, and so obtain the molecular radius on the supposition that the molecules may be regarded as elastic spheres.

The table on p. 183 gives the coefficients of viscosity of various gases, and the values of $\frac{1}{2}\sigma$, calculated from equation (210). These values are at least of the same order of magnitude as the values calculated from the deviations from Boyle's law in § 60. The reason why still better agreement cannot be expected will be clear from what has already been said in § 146.

Chapter VII

CONDUCTION OF HEAT

Elementary Theory

153. Chapter II contained a very incomplete discussion of the conduction of heat in a gas. We shall now attempt a more exact, although still imperfect, investigation of the problem.

The principle is that already explained. A molecule which describes a free path of length l with velocity C and total energy E is regarded as transporting energy E through a distance l, so that on balance there is a transport of energy from regions in which E is large to regions in which E is small— i.e. from places of high to places of low temperature.

Let a gas be supposed arranged in layers of equal temperature parallel to the plane of xy. Let $\bar{E}$ denote the mean energy of a molecule at any point in the gas, so that $\bar{E}$ will be a function of z.

Let us fix our attention on the molecules which cross a unit area of the plane $z = z_0$. Some molecules will cross this unit area after having come a distance l from their last collision in a direction making an angle θ with the axis of z. The last collision of these molecules must accordingly have taken place in the plane

$$z = z_0 - l \cos \theta.$$

We shall, for the moment, make the simplifying assumption that the mean energy of these molecules is that appropriate to this plane, and this may be taken to be

$$\bar{E} - l \cos \theta \frac{\partial \bar{E}}{\partial z}, \qquad \qquad \dots\dots(223)$$

where $\bar{E}$ is evaluated at $z = z_0$.

The number of molecules which cross the unit area in question in a direction making an angle between θ and $\theta + d\theta$ with the axis of z per unit time is (cf. formula (197))

$$\tfrac{1}{2} \nu \bar{c} \cos \theta \sin \theta \, d\theta,$$

and if we assume that each of these has an average amount of energy given by formula (223), the total flow of energy across the unit area of the plane will be

$$\int_{\theta=0}^{\theta=\pi} \left(\bar{E} - l \cos\theta \frac{\partial\bar{E}}{\partial z} \right) \tfrac{1}{2} v\bar{c} \cos\theta \sin\theta \, d\theta = -\tfrac{1}{3} v\bar{c}l \frac{\partial\bar{E}}{\partial z}.$$

$$\dots\dots(224)$$

If $\bar{E}$ had been independent of z, this flow of energy would of course have been *nil*, for as much would have crossed the plane in one direction as in the other. But if $\bar{E}$ increases with z, the molecules which cross the plane in the direction of z decreasing, carry more energy than those crossing the plane in the reverse direction, since they come from regions in which z is greater. Thus there is a resulting flow of energy in the direction of z decreasing.

If k is the coefficient of conduction of heat, the flow of heat across unit area of the plane $z = z_0$ in the direction of z increasing is $-k\dfrac{\partial T}{\partial z}$, so that the flow of energy is $-Jk\dfrac{\partial T}{\partial z}$, where J is the mechanical equivalent of heat.

Equating this to expression (224),

$$Jk\frac{\partial T}{\partial z} = \tfrac{1}{3}v\bar{c}l\frac{\partial\bar{E}}{\partial z} = \tfrac{1}{3}v\bar{c}l\frac{d\bar{E}}{dT}\frac{\partial T}{\partial z},$$

from which it follows that the value of k is

$$k = \frac{1}{3}\frac{v\bar{c}l}{J}\frac{d\bar{E}}{dT}. \qquad\dots\dots(225)$$

From equation (24) we have the relation

$$C_v = \frac{1}{Jm}\frac{d\bar{E}}{dT},$$

where C_v is the specific heat at constant volume, and again, from equation (203), if η is the coefficient of viscosity,

$$\eta = \tfrac{1}{3}v\bar{c}lm.$$

Using these relations, equation (225) becomes

$$k = \eta C_v. \qquad\dots\dots(226)$$

154. The flow of energy across the plane $z = z_0$ is at the rate of $-Jk\dfrac{\partial T}{\partial z}$ per unit area in the direction of z increasing. Across the plane $z = z_0 + dz$, the corresponding rate of flow is

$$-Jk\left(\frac{\partial T}{\partial z} + \frac{\partial^2 T}{\partial z^2}\,dz\right).$$

Consequently the slab of gas for which z lies between z_0 and $z_0 + dz$ gains energy at a rate

$$Jk\frac{\partial^2 T}{\partial z^2}\,dz \qquad\qquad \ldots\ldots(227)$$

per unit area, which must increase the temperature of the slab.

To raise the temperature at a rate $\dfrac{dT}{dt}$ requires energy per unit area equal to

$$J\rho\,C_v\frac{dT}{dt}.$$

Equating this to expression (227), we obtain

$$\rho\,C_v\frac{dT}{dt} = k\frac{\partial^2 T}{\partial z^2},$$

which is the ordinary Fourier equation of conduction of heat. Introducing the value of k from equation (226), this becomes

$$\frac{dT}{dt} = \frac{\eta}{\rho}\frac{\partial^2 T}{\partial z^2}, \qquad\qquad \ldots\ldots(228)$$

which is the special form of the equation of conduction appropriate to the kinetic theory.

Exact Theory

155. Equation (226) could not in any case be exact and detailed analysis shows it to be far from exact. There must clearly be an exact relation of the form

$$k = \epsilon\eta C_v,$$

where ϵ is a numerical multiplier, but the evaluation of ϵ presents a problem of great complexity. Many attempts have been made to evaluate ϵ by the approximate methods employed in the last chapter, but none of them has met with much success. For the special case of molecules repelling according to the inverse fifth power of the distance, Maxwell gave an exact theory which led to the value $\epsilon = 2\cdot5$ (see § 199 below). For monatomic gases Chapman* has evaluated ϵ by methods explained below (cf. § 187). In his first paper he obtained $\epsilon = 2\cdot500$ as a first approximation for all laws of force of the form μr^{-s}, this including of course the special case $s = 5$ studied by Maxwell. In his second paper he finds that further approximations alter this value by less than one per cent of its value. The greatest error in the first approximation is found to occur in the case of elastic spheres, for which the value of ϵ is $2\cdot522$.

Enskog[†] has obtained identical results, and has further given the general formula for monatomic molecules repelling as the inverse sth power of the distance

$$\epsilon = \frac{5}{2} \frac{1 + \dfrac{(s-5)^2}{4(s-1)(11s-13)} + \cdots}{1 + \dfrac{3(s-5)^2}{2(s-1)(101s-113)} + \cdots},$$

which reduces to Maxwell's exact value $\epsilon = \frac{5}{2}$ when $s = 5$.

EXPERIMENTAL VALUES

156. We proceed to examine the relation between k and η which is found experimentally.

Monatomic Gases. The following table of recent determinations of $k/\eta C_v$, the quantity we have denoted by ϵ, is given by Enskog:[‡]

Helium[§]	at	0° C.,	$\epsilon = 2\cdot40$,
	,,	$-191\cdot6$° C.,	$\epsilon = 2\cdot23$,
	,,	$-252\cdot1$° C.,	$\epsilon = 2\cdot02$.
Argon[§]	at	0° C.,	$\epsilon = 2\cdot49$,
	,,	$182\cdot5$° C.,	$\epsilon = 2\cdot57$.
Neon[‖]	at	10° C.,	$\epsilon = 2\cdot501$.

* *Phil. Trans. Roy. Soc.* **211** A (1912), p. 433 and **216** A (1915), p. 279.

† See footnote to p. 182.   ‡ *L.c.* p. 104.

§ Eucken, *Phys. Zeitschr.* **12** (1911), p. 1101; **14** (1913), p. 324.

‖ Bannawitz, *Ann. d. Phys.* **48** (1915), p. 577.

Other investigators have found similar, although not identical, values. Thus Schwarze** in 1903 found $\epsilon = 2\cdot507$†† for helium and $\epsilon = 2\cdot501$ for argon, while Hercus and Laby‡‡ give $\epsilon = 2\cdot31$ for helium and $\epsilon = 2\cdot47$ for argon. Thus the theoretical law seems to be confirmed to within the limits of experimental error except at low temperatures.

Other Gases. The following table contains, in its sixth column, some observed values of $k/\eta C_v$ for various gases. The values do not, in general, approximate either to $2\cdot5$ or to any other value.

Values of k and of $k/\eta C_v$

Gas	k (obs.)	Authority*	η (p. 183)	C_v	$k/\eta C_v$ (obs.)	$\frac{1}{4}(9\gamma - 5)$
Hydrogen	0·0003970	1	0·0000857	2·42	1·91	1·90
Helium	0·0003360	1	0·000188	0·746†	2·40	2·44
Carbon-monoxide	0·00005425	1	0·000166	0·177	1·85	1·91
Nitrogen	0·0000566	1	0·000167	0·178‡	1·91	1·91
Ethylene	0·0000407	1	0·0000961	0·274§	1·55	1·55
Air	0·0000566	1	0·000172	0·172‖	1·91	1·91
Nitric oxide	0·0000555	1	0·000179	0·167	1·86	1·88
Oxygen	0·0000570	1	0·000192	0·156	1·90	1·90
Argon	0·00003894	2	0·000210	0·0745¶	2·49	2·44
Carbon-dioxide	0·0000337	1	0·000137	0·156	1·58	1·72
Nitrous oxide	0·0000351	3	0·000137	0·148	1·73	1·73

* Authorities:

    1. Eucken, *Phys. Zeitschr.* **14** (1913), p. 324.

    2. Schwarze, *Ann. d. Phys.* **11** (1903), p. 303.

    3. Value assumed by Eucken (*l.c.*). This is the mean of determinations by Winkelmann and Wüllner.

† Determined by Vogel, and quoted by Eucken. The value of C_v for helium given by the formula (25) is, however, 0·767.

‡ Calculated from formula (25). Eucken takes $C_v = 0\cdot177$, Pier gives $C_v = 0\cdot175$.

§ The mean of values given by Winkelmann (*Pogg. Ann.* **159** (1876), p. 177) and Wüllner (*Wied. Ann.* **4** (1878), p. 321).

‖ Direct experimental value.

¶ The value assumed by Eucken. Schwarze uses the value $C_v = 0\cdot0740$, based upon an experimental determination of C_p by Dittenberger (*Halle Diss.* 1897). Pier gives $C_v = 0\cdot0746$. The theoretical value given by formula (25) is 0·767.

An inspection of the values obtained shews, however, that $k/\eta C_v$ is greatest for monatomic gases, and least for gases in which the molecules are of most complex structure (ethylene, carbon-

** *Ann. d. Phys.* **11** (1903), p. 303.

†† Eucken (*Phys. Zeitschr.* **14**, p. 328) states that Schwarze gives too high a value for helium owing to miscalculation of the value C_v.

‡‡ *Proc. Roy. Soc.* **95 A** (1918), p. 190.

dioxide, etc.). In other words $k/\eta C_v$ is largest when the specific heat C_v originates most in internal motions of the molecules. This leads us to suspect that when the molecules are regarded as carriers of energy, a distinction must be made between the energy of motion in space and the energy of internal motion. If we replace C_v by its value as found in § 21, equation (226) becomes

$$k = \eta C_v = \tfrac{3}{2}(1+\beta)\frac{R}{Jm}\eta. \qquad \ldots\ldots(229)$$

We have seen that when the energy is wholly translational ($\beta = 0$), the value of k given by this formula must be multiplied by (approximately) $\tfrac{5}{2}$. Eucken* has suggested that the simpler formula (229) may be accurate for the transport of internal energy, there being (cf. § 212, below) no correlation between the velocity of the molecule and the amount of internal energy carried.

Combining these two contributions to the transport of energy, we arrive at the formula

$$k = \tfrac{3}{2}(\tfrac{5}{2}+\beta)\frac{R}{Jm}\eta$$
$$= \frac{\tfrac{5}{2}+\beta}{1+\beta}\eta C_v$$
$$= \tfrac{1}{4}(9\gamma - 5)\eta C_v, \qquad \ldots\ldots(230)$$

the last two forms being obtained on substituting the values of C_v and γ from equations (29) and (30). According to this equation, ϵ, or $k/\eta C_v$, ought to have a value $\tfrac{1}{4}(9\gamma - 5)$ which depends only on γ, the ratio of the specific heats. In the last column of the table on p. 189 the values of $\tfrac{1}{4}(9\gamma - 5)$ are given, and are seen to be in fair agreement with the observed values of $k/\eta C_v$.

Conduction of Heat in Rarefied Gas

157. The theory of conduction of heat, like that of viscous flow, assumes a special form when the free path is larger than the dimensions of the apparatus. Molecules then transport heat across the whole apparatus at a single bound, so that the whole of the gas must be assumed to be at a uniform temperature;

---

* *Phys. Zeitschr.* **14** (1913), p. 324.

there is no longer a gradual temperature gradient. The appropriate mathematical theory has been developed by Knudsen.*

We consider a rarefied gas enclosed in a vessel, and fix our attention on a small area dS of the inner wall of the vessel.

If all the molecules of the gas moved with the same velocity c, the number of molecules impinging on the area dS in unit time would be, as in § 39,

$$\tfrac{1}{4}\nu c \, dS.$$

Each of these would deliver up energy $\tfrac{1}{2}mc^2$ to the wall, so that the total energy transferred from the gas to the element dS would be

$$\tfrac{1}{8}\rho c^3 \, dS.$$

If the molecules do not all move with the same speed, we must average c^3 in this expression. If the speeds are distributed according to Maxwell's law, we find (cf. Appendix VI, p. 306) that the average value of c^3 is $4/\sqrt{(\pi h^3 m^3)}$, and the total transfer of energy from gas to wall, per unit area and unit time, becomes

$$\frac{1}{2}\frac{\rho}{\sqrt{\pi h^3 m^3}} = \tfrac{1}{2}\rho\left(\frac{2RT}{\pi m}\right)^{\tfrac{1}{2}}\left(\frac{2RT}{m}\right) = \tfrac{1}{2}p\bar{c},$$

where p is the pressure in the gas.

If the wall is at the same temperature as the gas, the net transfer of energy between the wall and the gas must be nil, so that the wall must yield back energy to the gas at a rate $\tfrac{1}{2}p\bar{c}$. Suppose however that the gas is at some temperature T_0 (absolute), while the wall is at some other temperature T_1.

The simplest tentative assumption to make is that while the molecules of the gas are in their condition of adsorption by the wall (§ 34), they acquire the mean energy which corrresponds to the temperature of the wall, and subsequently leave the wall with this mean energy. The molecules will now take away from the wall T_1/T_0 times the amount of energy they took to it, so that there will be a net transfer of energy (still measured

---

* *Ann. d. Physik,* **31** (1910), p. 205, **33** (1910), p. 1435 and later papers. See also M. Knudsen, *The Kinetic Theory of Gases* (Methuen, 1934), p. 46, and Loeb, *Kinetic Theory of Gases* (2nd ed., 1934), pp. 325 ff.

per unit area per unit time) from the wall to the gas of amount

$$\tfrac{1}{2}p\bar{c}\left(\frac{T_1}{T_0}-1\right) = \frac{1}{2}\frac{p\bar{c}(T_1-T_0)}{T_0}. \qquad \ldots\ldots(231)$$

Thus the heat dissipated through contact with a rarefied gas will be jointly proportional to the temperature difference $T_1 - T_0$ and to the pressure p.

158. Experiment confirms this law of proportionality, but generally speaking, does not confirm the multiplying factor in formula (231). The actual transfer of heat is substantially less than that predicted by formula (231), as though the interchange of energy between the solid wall and the adsorbed molecules of gas were far from complete, an explanation suggested by v. Smoluchowski[*] in 1898, and again by Soddy and Berry[†] in 1910.

This led Knudsen[‡] to introduce a quantity a which he described as a "coefficient of accommodation". He imagines, in brief, that when a molecule is adsorbed by the wall, its energy is not adjusted through the whole range of temperature difference $T_1 - T_0$, but only through a fraction a of this range. Formula (231) must now be replaced by

$$\frac{1}{2}\frac{p\bar{c}(T_1-T_0)}{T_0}a, \qquad \ldots\ldots(232)$$

and it is immediately possible to determine a by experiments on the dissipation of heat.

The value of a is found to depend very largely on the temperature and other physical conditions, such as, in particular, the cleanness or otherwise of the solid surface. Almost all values for a from 0 to 1 appear to be possible. Thus for CO_2 in contact with platinum heavily coated with platinum black, Knudsen found $a = 0.975$; for helium in contact with tungsten, J. K. Roberts[§] found that a was equal to 0.057 at $22°$ C., and fell steadily to 0.025 as the temperature was lowered to $-194°$ C. A

---

[*] M. v. Smoluchowski, *Wied. Ann.* **64** (1898), p. 101.

[†] *Proc. Roy. Soc.* **83** A (1910), p. 254 and **84** A (1911), p. 576.

[‡] *Ann. d. Phys.* **34** (1911), p. 593, **36** (1911), p. 871, and **6** (1930), p. 129. For a full discussion of the subject see Loeb, *Kinetic Theory of Gases* (2nd ed., 1934), pp. 321 ff.

[§] *Proc. Roy. Soc.* **129** A (1930), p. 146 and **135** A (1932), p. 192.

study of the relation between a and the temperature suggested that a would be found to vanish at the absolute zero.

General dynamical considerations suggest that the value of a should be substantially less than unity. In § 10 we considered the impact of gas molecules of mass m' on wall molecules of mass m, both molecules being treated as hard elastic spheres obeying the Newtonian laws of motion. We found that the average impact resulted in the gas gaining energy of amount

$$\frac{2mm'}{(m+m')^2}(\overline{mu^2} - \overline{m'u'^2}).$$

This is equal to
$$\frac{4mm'}{(m+m')^2}$$

times the interchange of energy which would occur if the gas molecules took up the temperature of the wall completely, and so corresponds to a "coefficient of accommodation" a given by

$$a = \frac{4mm'}{(m+m')^2} = 1 - \left(\frac{m-m'}{m+m'}\right)^2.$$

This value of a is always less than unity, and is substantially less except when the masses m, m' are nearly equal. For more complicated molecules and more complicated impacts, the theoretical value of a would no doubt be very different from the simple value just found. Various attempts have been made to calculate a "coefficient of accommodation" which shall agree with experiment, and from these it seems to emerge quite clearly that the problem is one for wave-mechanics, and so is outside the scope of the present book.* Jackson and Howarth† have found a wave-mechanics formula which represents the observations of Roberts mentioned above with very fair accuracy.

159. Our theoretical calculations have been based on the supposition that the only energy which the impinging molecule can transfer to the wall is its kinetic energy of motion $\frac{1}{2}mc^2$. We have already seen (§ 21) that many molecules have other energy besides this, and the question arises as to what happens to this

* J. M. Jackson, *Proc. Camb. Phil. Soc.* **28** (1932), p. 136; C. Zener, *Phys. Review*, **37** (1931), p. 557 and **40** (1932), pp. 178, 335; Jackson and Mott, *Proc. Roy. Soc.* **137** A (1932), p. 703.

† *Proc. Roy. Soc.* **142** A (1933), p. 447.

other energy when the molecule impinges on a solid. Knudsen*
has devised ingenious methods for probing this question experimentally and has reached the conclusion that the internal
molecular energy also has an accommodation coefficient, which,
to within the limits of experimental error, is the same as the
accommodation coefficient for the kinetic energy of translation.
Thus if the internal energy is β times the kinetic energy of
rotation, the total transfer of energy is $(1+\beta)$ times that given
by formula (232).

CONDUCTION OF HEAT AND ELECTRICITY IN SOLIDS

Conduction of Heat

160. In 1900 Drude† propounded a theory of conduction of
heat in solids, according to which the process is exactly similar to
that in gases which we have just been considering, except that
the carriers of the heat-energy are the free electrons in the metals.

According to the simplest form of this theory, the coefficient
of conduction of heat in a solid will be given by equation (225),
namely

$$k = \frac{1}{3}\frac{\bar{v}\bar{c}l}{J}\frac{d\bar{E}}{dT}, \qquad \ldots\ldots(233)$$

in which all the quantities refer to the free electrons in the solid,
so that ν is the number of free electrons per unit volume, l is their
average free path as they thread their way through the solid, and
so on. Taking $\bar{E} = \frac{3}{2}RT$, this becomes

$$k = \frac{1}{2J}\bar{v}\bar{c}lR. \qquad \ldots\ldots(234)$$

Conduction of Electricity

161. Drude's theory supposes that the free electrons also act
as carriers in the conduction of electricity. If there is an electric
force $\varXi$ in the direction of the axis of x, each electron will be acted
on by a force $\varXi e$, and so will gain momentum in the direction Ox
at a rate $\varXi e$ per unit time. The time required to describe an average

* *Ann. d. Phys.* **34** (1911), p. 593.
† *Ann. d. Phys.* **1** (1900), p. 566.

free path l, with average velocity $\bar{c}$, will be $l/\bar{c}$, so that in describing such a free path, the electron will acquire an additional momentum in the direction of the axis of x equal to $\Xi el/\bar{c}$.

Since the mass of the electron is very small compared with that of the atom or molecule with which it collides, we must suppose (cf. § 130) that there is no persistence of velocities after collisions, so that an electron starts out from collision with a velocity for which all directions are equally likely, and, in describing its free path, superposes on to this a velocity

$$\frac{\Xi el}{m\bar{c}}$$

parallel to the axis of x. It follows that at any instant the free electrons have an average velocity u_0, parallel to the axis of x, given by

$$u_0 = \frac{1}{2}\frac{\Xi el}{m\bar{c}}.$$

Across unit area perpendicular to the axis of x, there will be a flow of electrons at the rate νu_0 per unit time, and these will carry a current i given by

$$i = \nu e u_0 = \frac{1}{2}\frac{\Xi \nu e^2 l}{m\bar{c}}.$$

The coefficient of electric conductivity σ is defined by the relation $i = \sigma \Xi$, and is therefore equal to the coefficient of Ξ in the above equation. To the order of accuracy to which we are now working, $\bar{c}$ may be supposed to be the velocity of each electron, so that we may put $\frac{1}{2}m(\bar{c})^2 = \frac{3}{2}RT$, and the conductivity is given by

$$\sigma = \frac{\nu e^2 l\bar{c}}{6RT}. \qquad \qquad \text{......(235)}$$

This is Drude's formula for electric conductivity.

Ratio of the two Conductivities

162. *The Wiedemann-Franz law.* By comparison of equations (234) and (235), we obtain

$$\frac{k}{\sigma} = 3\left(\frac{R}{e}\right)^2\frac{T}{J},$$

which is Drude's approximate formula for k/σ. From this equation it appears that:

> at a given temperature, the ratio of the electric and thermal conductivities must be the same for all substances.

This is the law of Wiedemann and Franz, announced by them as an empirical discovery in 1853.[*]

Various attempts have been made to obtain a more exact formula for k/σ. In 1911, N. Bohr[†] obtained the formula

$$\frac{k}{\sigma} = \frac{2s}{s-1}\left(\frac{R}{e}\right)^2 \frac{T}{J}, \qquad \dots\dots(236)$$

in which the electrons are supposed to be repelled from the atoms according to the law μr^{-s}. According to this, the law of Wiedemann and Franz is not strictly true, but k/σ varies only through the factor $2s/(s-1)$ and as s varies from $s = 5$ to $s = \infty$, this only varies from 2·5 to 2.

The Law of Lorenz. From equation (236) it also follows that:

> the ratio of the thermal and electric conductivities must be proportional to the absolute temperature.

a law put forward on theoretical grounds by Lorenz in 1872.[‡]

Comparison with Experiment

163. For elastic spheres ($s = \infty$) equation (236) reduces to

$$\frac{k}{\sigma} = 2\left(\frac{R}{e}\right)^2 \frac{T}{J}, \qquad \dots\dots(237)$$

a formula originally given by Lorentz.[§] Inserting numerical values, this equation becomes

$$\frac{k}{\sigma T} = 3\cdot538 \text{ in electromagnetic units.}$$

Extensive experiments to test this formula have been made by Jäger and Diesselhorst[||], C. H. Lees[¶] and many others. In general these confirmed the theoretical equations, except that

---

[*] *Pogg. Ann.* **89** (1853), p. 497.

[†] *Studier over Metallernes Elektrontheorie* (Copenhagen, 1911).

[‡] *Pogg. Ann.* **147** (1872), p. 429 and *Wied. Ann.* **13** (1882), p. 422.

[§] *The Theory of Electrons*, p. 67 and note 29.

[||] *Berlin. Sitzungsber.* **38** (1899), p. 719, and *Abhand. d. Phys.-Tech. Reichsanstalt*, **3** (1900), p. 369.

[¶] C. H. Lees, "The effects of low temperatures on the Thermal and Electrical conductivities of certain approximately pure metals and alloys", *Phil. Trans.* **208** A (1908), p. 381.

the observed values of k/σ were always somewhat too high. Jäger and Diesselhorst experimented at 18° C. and at 100° C. Putting $T = 291\cdot2$ (i.e. 18° C.) in formula (235), and measuring k in work, instead of in heat, units (i.e. omitting the J in the denominator), formula (235) gives

$$\frac{k}{\sigma} = 4\cdot31 \times 10^{10} \text{ at } 18° \text{ C.}$$

The simpler formula of Drude leads to $\frac{3}{2}$ times this value, namely

$$\frac{k}{\sigma} = 6\cdot46 \times 10^{10} \text{ at } 18° \text{ C.}$$

The following are examples of the results obtained experimentally by Jäger and Diesselhorst:

for three samples of copper, $k/\sigma = 6\cdot76,\ 6\cdot65,\ 6\cdot71 \times 10^{10}$,

for silver, $\qquad\qquad\qquad k/\sigma = 6\cdot86 \times 10^{10}$,

for two samples of gold, $\quad k/\sigma = 7\cdot27,\ 7\cdot09 \times 10^{10}$.

As temperature coefficient of this ratio they find:

for two samples of copper, 0·39, 0·39 per cent,

for silver, $\qquad\qquad\qquad$ 0·37 per cent,

for two samples of gold, $\quad$ 0·36, 0·37 per cent.

The theoretical value, as given by equation (237), is 0·366 per cent. Lees finds that there is a greater divergence from formula (237) as the temperature of liquid air is approached.

It is more difficult to test the theoretical values for k and σ separately, since the formulae for these coefficients separately contain the quantities ν and l, for which it is difficult to form a reliable numerical estimate. But such evidence as is available shews quite definitely that the formulae for k and σ separately do not shew anything like so good an agreement with observation as that shewn by the formula for their ratio k/σ.

Indeed the phenomenon of electric super-conductivity shews that this must be the case. At helium temperatures the resistance may be only a fraction 10^{-11} times the resistance at ordinary temperatures, but it is impossible to believe that at these temperatures the free path can suddenly increase its length by a factor 10^{11}.

Chapter VIII

DIFFUSION

ELEMENTARY THEORIES

164. The difficulties in the way of an exact mathematical treatment of diffusion are similar to those which occurred in the problems of viscosity and heat conduction. Following the procedure we adopted in discussing these earlier problems, we shall begin by giving a simple, but mathematically inexact, treatment of the question.

We imagine two gases diffusing through one another in a direction parallel to the axis of z, the motion being the same at all points in a plane perpendicular to the axis of z. The gases are accordingly arranged in layers perpendicular to this axis.

The simplest case arises when the molecules of the two gases are similar in mass and size—like the red and white billiard balls we discussed in § 6. In other cases differences in the mass and size of the molecules tend, as the motion of the molecules proceeds, to set up differences of pressure in the gas. The gas adjusts itself against these by a slow mass-motion, which will of course be along the axis of z at every point.

Let us denote the mass-velocity in the direction of z increasing by w_0, and let the molecular densities of the two gases be ν_1, ν_2. Then ν_1, ν_2 and w_0 are functions of z only.

We assume that, to the approximation required in the problem, the mass-velocity of the gas is small compared with its molecular-velocity, and we also assume that the proportions of the mixture do not change appreciably within distances comparable with the average mean free path of a molecule. We shall also, to obtain a rough first approximation, assume that Maxwell's law of distribution of velocities obtains at every point, and that h is the same for the two gases.

165. The number of molecules of the first kind, which cross the plane $z = z_0$ per unit area per unit time in the direction of z increasing, is now

$$\left(\frac{hm_1}{\pi}\right)^{\frac{3}{2}} \iiint \nu_1 e^{-hm_1[u^2+v^2+(w-w_0)^2]} w\, du\, dv\, dw, \quad \ldots\ldots(238)$$

in which the limits are from $-\infty$ to $+\infty$ as regards u and v, and from 0 to ∞ as regards w.

In accordance with the principles already explained, ν must be evaluated at the point from which the molecules started after their last collision. Those which move so as to make an angle θ with the axis of z may be supposed, on the average, to come from a point of which the z coordinate is $z_0 - \lambda \cos \theta$, and at this point the value of ν_1 may be taken to be

$$\nu_1 = (\nu_1)_{z=z_0} - \lambda \cos \theta \left(\frac{\partial \nu_1}{\partial z}\right). \quad \ldots\ldots(239)$$

Inserting this value for ν_1 into expression (238), this expression becomes the difference of two integrals. The first is expression (238) with ν_1 taken outside the signs of integration and evaluated at $z = z_0$. The value of this integral is easily found to be

$$\tfrac{1}{2}(\nu_1)_{z=z_0}(\tfrac{1}{2}\bar{c}_1 + w_0), \quad \ldots\ldots(240)$$

where $\bar{c}_1$ denotes the mean molecular-velocity of all the molecules of the first kind.

The second integral is

$$\lambda\left(\frac{\partial \nu_1}{\partial z}\right)\left(\frac{hm_1}{\pi}\right)^{\frac{3}{2}} \iiint e^{-hm_1[u^2+v^2+(w-w_0)^2]} w \cos \theta\, du\, dv\, dw.$$
$$\ldots\ldots(241)$$

Owing to the presence of the multiplier $\lambda\left(\dfrac{\partial \nu_1}{\partial z}\right)$, this expression is already a small quantity of the first order, so that in evaluating it we may put $w_0 = 0$. Replacing $\cos \theta$ by w/c, it becomes

$$\lambda\left(\frac{\partial \nu_1}{\partial z}\right)\left(\frac{hm_1}{\pi}\right)^{\frac{3}{2}} \iiint e^{-hm_1 c^2} \frac{w^2}{c}\, du\, dv\, dw,$$

in which the integral is taken over all values of u and v, and over all positive values of w.

This expression is easily evaluated by noticing that it has just half the value it would have if taken over all values of u, v, w, and is therefore equal to $\frac{1}{2}\lambda\left(\frac{\partial \nu_1}{\partial z}\right)$ times the average value of $\frac{w^2}{c}$ taken over all molecules, in a gas having no mass-motion. This average value is equal to one-third of the average of $\frac{u^2+v^2+w^2}{c}$ or c, and is therefore $\frac{1}{3}\bar{c}_1$. Hence expression (241) is equal to

$$\frac{1}{6}\lambda\left(\frac{\partial \nu_1}{\partial z}\right)\bar{c}_1.$$

Combining this with expression (240), we find as the total value of expression (238),

$$\frac{1}{2}\nu_1(\frac{1}{2}\bar{c}_1+w_0)-\frac{1}{6}\lambda\frac{\partial \nu_1}{\partial z}\bar{c}_1,$$

in which all quantities are to be evaluated in the plane $z = z_0$.

This is the total flow of molecules across unit area of the plane $z = z_0$, in the direction of z increasing. The corresponding flow in the opposite direction is

$$\frac{1}{2}\nu_1(\frac{1}{2}\bar{c}_1-w_0)+\frac{1}{6}\lambda\frac{\partial \nu_1}{\partial z}\bar{c}_1.$$

The rate of increase of the number of molecules of the first kind on the positive side of the plane $z = z_0$, measured per unit time per unit area, is the difference of these two expressions. Denoting this quantity by Γ_1, we have

$$\Gamma_1 = \nu_1 w_0 - \frac{1}{3}\lambda_1\frac{\partial \nu_1}{\partial z}\bar{c}_1.$$

Similarly, for the rate of increase of molecules of the second kind,

$$\Gamma_2 = \nu_2 w_0 - \frac{1}{3}\lambda_2\frac{\partial \nu_2}{\partial z}\bar{c}_2.$$

If the flow is to be steady, the total flow of molecules over every plane must be zero, and this requires that

$$\Gamma_1+\Gamma_2 = 0.$$

Further, the pressure must be constant throughout the gas, so that we must have

$$\nu_1+\nu_2 = \text{cons.},$$

whence, by differentiation with respect to z,

$$\frac{\partial \nu_1}{\partial z} + \frac{\partial \nu_2}{\partial z} = 0.$$

Eliminating w_0 from these equations, we now obtain

$$-\Gamma_1 = \Gamma_2 = \frac{1}{3}\frac{\nu_1 \lambda_2 \bar{c}_2 + \nu_2 \lambda_1 \bar{c}_1}{\nu_1 + \nu_2}\frac{\partial \nu_1}{\partial z}. \qquad \ldots\ldots(242)$$

The number of molecules of the first kind, in a layer of unit cross-section between the planes $z = z_0$ and $z = z_0 + dz$, is $\nu_1 dz$; the rate at which this quantity increases is $\dfrac{d\nu_1}{dt}dz$, but is also found to be $-\dfrac{\partial \Gamma_1}{\partial z}dz$, by calculating the flow across the two boundary planes. Hence we have

$$\frac{d\nu_1}{dt} = -\frac{\partial \Gamma_1}{\partial z},$$

and on using the value of Γ_1 just found, and neglecting small quantities of the second order, this becomes

$$\frac{d\nu_1}{dt} = \mathfrak{D}_{12}\frac{\partial^2 \nu_1}{\partial z^2}, \qquad \ldots\ldots(243)$$

where $$\mathfrak{D}_{12} = \frac{1}{3}\frac{\nu_1 \lambda_2 \bar{c}_2 + \nu_2 \lambda_1 \bar{c}_1}{\nu_1 + \nu_2}. \qquad \ldots\ldots(244)$$

Equation (243) is the well-known equation of diffusion, $\mathfrak{D}_{12}$ being the coefficient of diffusion of the two gases. Hence the coefficient of diffusion is given by formula (244). It is symmetrical as regards the physical properties of the two gases, but depends on the ratio ν_1/ν_2 in which they are mixed.

The foregoing analysis is essentially the same as that given by Meyer in his *Kinetic Theory of Gases*, and formula (244) is generally known as Meyer's formula* for the coefficient of diffusion.

* The actual value of $\mathfrak{D}_{12}$ given by Meyer (*Kinetic Theory of Gases*, p. 255, English trans.) is $\frac{3}{8}\pi$ times that given by formula (244). Meyer's formula has, however, attempted to take into account a correction which is here reserved for later discussion (§ 168). Meyer does not claim that his correction is exact.

Coefficient of Self-diffusion

166. Formula (244) becomes especially simple when the molecules of the two gases are approximately of equal size and weight. If we agree to neglect the differences in size and weight, we may take λ and c to be the same for each gas, and so obtain

$$\mathfrak{D} = \tfrac{1}{3}\lambda\bar{c}. \qquad \ldots\ldots(245)$$

Comparing this with the corresponding approximate formula (203) for the coefficient of viscosity

$$\eta = \tfrac{1}{3}\lambda\bar{c}\rho,$$

we obtain the relation $\qquad \mathfrak{D} = \dfrac{\eta}{\rho}. \qquad \ldots\ldots(246)$

The quantity $\mathfrak{D}$ obtained in this way may also be regarded as the coefficient of self-diffusion or interdiffusivity of a single gas. It measures the rate at which selected molecules of a homogeneous gas diffuse into the remainder, as for instance red billiard balls into white billiard balls in our analogy of § 6, or (approximately) one isotope into another isotope of the same element. Hartek and Schmidt have made diffusion experiments with normal hydrogen and a mixture rich in para-hydrogen, while Boardman and Wild* have experimented with mixtures of gases—as for instance nitrogen and carbon-monoxide—in which the molecules are not only of equal mass, but are believed to be of very similar structure (see § 176).

Dependence on Proportions of Mixture

167. In the special case just considered, the value of $\mathfrak{D}$ is independent of the proportion of the mixture, but in the more general case formula (244) shews $\mathfrak{D}_{12}$ ought to vary with the proportions of the mixture. In the limiting case in which $\nu_1/\nu_2 = 0$,

$$\mathfrak{D}_{12} = \tfrac{1}{3}\lambda_1\bar{c}_1 = \frac{2}{3\sqrt{\pi h(m_1+m_2)}\,\pi\nu S_{12}^2}\left(\frac{m_2}{m_1}\right)^{\!\frac{1}{2}},$$

and there is a similar formula for the case of $\nu_2/\nu_1 = 0$, in which m_1 and m_2 are interchanged.

---

* *Proc. Roy. Soc.* **162** A (1937), p. 516.

Thus the coefficients of diffusion in these two cases stand in the ratio

$$\frac{\mathfrak{D}_{\nu_1=0}}{\mathfrak{D}_{\nu_2=0}} = \frac{m_2}{m_1}, \qquad \dots\dots(247)$$

shewing that the value of $\mathfrak{D}$ ought according to Meyer's formula to vary with the proportions of the mixture to the extent to which m_1 differs from m_2. For example, for the diffusion of H_2—CO_2, the extreme variation would be 22 to 1, for A—He it would be 10 to 1, and so on.

We shall see later that the observed variation of $\mathfrak{D}_{12}$ with ν_1/ν_2 is nothing like as great as is predicted by this formula. For the moment we proceed to correct the formulae for persistence of velocities, and shall find that the corrected equations predict a much smaller dependence of $\mathfrak{D}_{12}$ on ν_1/ν_2.

Correction to Meyer's Theory when the Molecules are Elastic Spheres

168. We shall consider first the correction to be applied to the simple formula for self-diffusion, namely

$$\mathfrak{D} = \tfrac{1}{3}\lambda\bar{c} \qquad \dots\dots(248)$$

$$= \frac{\eta}{\rho}. \qquad \dots\dots(249)$$

Actually, two sources of error have been introduced into these approximate equations, the first arising from the assumption that λ is the same for all velocities, and the second from neglect of the persistence of velocities.

As regards the first, λ must simply be replaced by λ_c and taken under the sign of integration in expression (241). The upshot of this is, that instead of λ in the final result, we have l, where

$$l = \frac{\displaystyle\int_0^\infty \lambda_c e^{-hmc^2} c^3 \, dc}{\displaystyle\int_0^\infty e^{-hmc^2} c^2 \, dc} = \frac{\overline{\lambda_c c}}{\bar{c}}.$$

This however is exactly the same as the l of the viscosity formula, so that this correction affects $\mathfrak{D}$ and κ exactly similarly, multiplying each by 1·051, but does not affect equation (249).

169. There remains the correction for the persistence of velocities. We found in § 139, that when a molecule arrives at the plane $z = z_0$ in a given direction, the expectation of the distance it has travelled in that direction is not λ, but $k\lambda$, where

$$k = \frac{1}{1-\theta}.$$

Here θ is the persistence of velocities at a collision between two molecules of equal mass, of which the value was found in § 126 to be 0·406. Thus the expectation of the molecule belonging to the one gas or the other is not that appropriate to a distance λ back, but to a distance $k\lambda$, and the effect of "persistence" is therefore to multiply the value of $\mathfrak{D}$ given in equation (249) by a factor k. Also, as we saw in § 139, the effect of persistence on the coefficient of viscosity is to multiply the simple expression $\frac{1}{3}\lambda\bar{c}\rho$ by a factor $1/(1-\frac{1}{2}\theta)$.

The values of $\mathfrak{D}$ and η, both corrected for persistence, are accordingly

$$\mathfrak{D} = \frac{1}{3(1-\theta)}\lambda\bar{c},$$

$$\eta = \frac{1}{3(1-\frac{1}{2}\theta)}\lambda\bar{c}\rho,$$

and the corrected form of equation (249) must be

$$\mathfrak{D} = \frac{1-\frac{1}{2}\theta}{1-\theta}\frac{\eta}{\rho}.$$

Putting $\theta = 0\cdot406$, the value found in § 126, this becomes

$$\mathfrak{D} = 1\cdot342\frac{\eta}{\rho}. \qquad \ldots\ldots(250)$$

It is of interest to examine into the origin of the difference between the effect of persistence of velocities on diffusion on the one hand, and on viscosity and conduction of heat on the other. Diffusion, it will be seen, is a transport of a *quality*, while viscosity and heat-conduction are transports of *quantities*. The difference rests ultimately upon the circumstance that qualities remain unaltered by collisions, whereas quantities do not.

When the molecules are not of equal mass, it is more difficult to estimate the effect of persistence.

When the molecules were equal, the expectation of the distance a molecule had come was increased by persistence from λ to

$$\lambda + \theta\lambda + \theta^2\lambda + \theta^3\lambda + \ldots = \frac{\lambda}{1-\theta}. \qquad \ldots\ldots(251)$$

When the molecules are of unequal masses, the persistence is different at different collisions, and instead of expression (251), we shall have one of the form

$$\lambda + p\lambda + pq\lambda + pqr\lambda + \ldots, \qquad \ldots\ldots(252)$$

where $p, q, r, \ldots$ are the different persistences at the various collisions. Suppose we are considering the motion of a molecule of mass m_1 in a mixture of molecules of masses m_1, m_2 mixed in the proportion ν_1/ν_2. Then of the quantities $p, q, r, \ldots$ a certain proportion, say β, of the whole will have an average value $\theta = 0\cdot406$, these representing collisions with other molecules of the first kind, while the remainder, a proportion $1 - \beta$ of the whole, will have an average value which we shall denote by θ_{12}, this being the persistence for a molecule of the first kind colliding with one of the second kind.

Let P denote expression (255), and let s denote $\beta\theta + (1 - \beta)\theta_{12}$, this being the expectation of each of the quantities $p, q, r, \ldots$. We have

$$P = \lambda + p\lambda + pq\lambda + pqr\lambda + \ldots,$$
$$Ps = \quad s\lambda + ps\lambda + pqs\lambda + \ldots,$$

and hence, by subtraction,

$$P(1-s) = \lambda + (p-s)\lambda + p(q-s)\lambda + pq(r-s)\lambda + \ldots.$$

Clearly the expectation of the right-hand side is λ, for the expectations of $p - s, \ q - s, \ r - s, \ldots$ are all zero. Hence the expectation of P is

$$P = \frac{\lambda}{1-s} = \frac{\lambda}{1 - \{\beta\theta + (1-\beta)\theta_{12}\}}. \qquad \ldots\ldots(253)$$

Accordingly the effect of persistence in this mixture of gases is to increase λ to a value of which the expectation is that on the right-hand side of equation (253).

170. In § 111 we found for the mean chance of collision per unit time for a molecule of the first kind, moving in a mixture of two kinds of gas,

$$2\nu_1\sigma_1^2\sqrt{\frac{2\pi}{hm_1}} + 2\nu_2 S_{12}^2\sqrt{\frac{\pi}{h}\left(\frac{1}{m_1}+\frac{1}{m_2}\right)}.$$

In this the first term results from collisions with molecules of the first kind, and the second from collisions with molecules of the second kind. These two terms must therefore stand in the ratio $\beta : 1 - \beta$, so that

$$\frac{\beta}{\sqrt{2}\pi\nu_1\sigma_1^2} = \frac{1-\beta}{\pi\nu_2 S_{12}^2\sqrt{1+\dfrac{m_1}{m_2}}} = \frac{1}{\sqrt{2}\pi\nu_1\sigma_1^2 + \pi\nu_2 S_{12}^2\sqrt{1+\dfrac{m_1}{m_2}}},$$

which again is equal to λ_1, the free path of a molecule of the first kind, by equation (151).

Using this value for β, we find for the value of expression (253), the free path of a molecule of the first kind increased by persistence,

$$P_1 = \frac{1}{(1-\theta)\sqrt{2}\pi\nu_1\sigma_1^2 + (1-\theta_{12})\pi\nu_2 S_{12}^2\sqrt{1+\dfrac{m_1}{m_2}}},$$

$$\dots\dots(254)$$

and there is a corresponding quantity P_2 for the second molecule.

On replacing λ_1, λ_2 in equation (244) by their enhanced values, as just found, we obtain as the form of Meyer's equation, after correction for persistence of velocities,

$$\mathfrak{D}_{12} = \frac{1}{3}\frac{\nu_1 P_2 \bar{c}_2 + \nu_2 P_1 \bar{c}_1}{\nu_1 + \nu_2}. \qquad \dots\dots(255)$$

In this formula the value of θ is always $0\cdot406$, while the value of θ_{12} depends, as was seen in § 128, on the ratio of the two masses. It was there found that θ_{12} was of the form

$$\theta_{12} = \frac{m_1 - \alpha_{12}m_2}{m_1 + m_2},$$

where α_{12} was a small positive number, depending on the ratio of the masses, but lying always between 0 and $\frac{1}{3}$, and equal to $0\cdot188$ for equal masses.

When ν_1 is small, we have, instead of the limiting form given in §167,

$$\mathfrak{D}_{12} = \frac{1}{3}\frac{\nu_2 P_1 \bar{c}_1}{\nu_1+\nu_2} = \frac{2}{3(1+\alpha_{12})\,\pi(\nu_1+\nu_2)\,S_{12}^2}\sqrt{\frac{1}{\pi h}\left(\frac{1}{m_1}+\frac{1}{m_2}\right)}.$$
$$\ldots\ldots(256)$$

When ν_2 is small, the limiting form is the same except that α_{12} is replaced by α_{21}. Thus the ratio of the extreme values of $\mathfrak{D}$ as ν_1/ν_2 varies is

$$\frac{\mathfrak{D}_{\nu_1=0}}{\mathfrak{D}_{\nu_2=0}} = \frac{1+\alpha_{21}}{1+\alpha_{12}},$$

instead of the ratio $m_2 : m_1$ found from Meyer's formula (247). Since the extreme values possible for α are 0 and $\frac{1}{3}$, it appears that the greatest range possible for $\mathfrak{D}$ is at most one of $4 : 3$.

Thus, when persistence of velocities is taken into account, Meyer's formula yields values which do not vary greatly with the proportion $\nu_1 : \nu_2$ of the mixture.*

The Stefan-Maxwell Theory

171. Equation (243) is of the same form as the well-known equation of conduction of heat: it indicates a progress or spreading out of the gas of the first kind, similar to the progress and spreading out of heat in a problem of conduction. The larger $\mathfrak{D}$ is, the more rapidly this progress takes place; $\mathfrak{D}$ is largest when the free paths are longest, and vice versa. Long free paths mean rapid diffusion, as we should expect.

Now the formula for the mean path λ_1 in a mixture of two gases was found in §111 to be

$$\lambda_1 = \frac{1}{\sqrt{2}\pi\nu_1\sigma_1^2 + \sqrt{\left(1+\dfrac{m_1}{m_2}\right)\pi\nu_2 S_{12}^2}}, \qquad \ldots\ldots(257)$$

* This was pointed out in a valuable paper by Kuenen ("The diffusion of Gases according to O. E. Meyer", Supp. 28 to the *Comm. Phys. Lab. Leiden*, Jan. 1913). Kuenen took α uniformly equal to 0·188, its value when $m_1 = m_2$, and assumed the number of collisions to be in the ratio $\nu_1^2 : \nu_1\nu_2$, so that his result is different from mine, but the principle was essentially the same. In a later paper by the same author (*Comm. Phys. Lab. Leiden*, Supp. 38) the mass difference was taken into account.

where S_{12} is the arithmetic mean of the diameters of the two kinds of molecules. The larger the denominator in this expression, the smaller λ_1 will be, and so the slower the process of diffusion. Both terms in the denominator of expression (257) accordingly contribute something towards hindering the process of diffusion.

The second of these terms arises from collisions of the molecules of the first kind with molecules of the second kind, and that these collisions should hinder diffusion is intelligible enough. But it is not so clear how collisions of the molecules of the first kind with one another, represented by the first term in the denominator of expression (257), can hinder the process of diffusion. When molecules of the same kind collide, their average forward motion remains unaffected by the conservation of momentum, and it is not easy to see how the process of diffusion has been hindered.

Stefan* and Maxwell† have accordingly suggested that collisions between molecules of the same kind should be entirely disregarded; we then have effective free paths given by the equations

$$\lambda_1 = \frac{1}{\sqrt{\left(1+\dfrac{m_1}{m_2}\right)\pi\nu_2 S_{12}^2}}; \quad \lambda_2 = \frac{1}{\sqrt{\left(1+\dfrac{m_2}{m_1}\right)\pi\nu_1 S_{12}^2}}$$

$$\dots\dots(258)$$

in place of equation (257). Using these values for the free paths, equation (244) becomes

$$\mathfrak{D}_{12} = \frac{m_1^{\frac{1}{2}}\bar{c}_2 + m_2^{\frac{1}{2}}\bar{c}_1}{3\pi(\nu_1+\nu_2)S_{12}^2(m_1+m_2)^{\frac{1}{2}}}$$

$$= \frac{2}{3\pi(\nu_1+\nu_2)S_{12}^2}\sqrt{\frac{1}{\pi h}\left(\frac{1}{m_1}+\frac{1}{m_2}\right)}, \qquad \dots\dots(259)$$

---

* *Wien. Sitzungsber.* **63** [2] (1871), p. 63, and **65** (1872), p. 323.

† *Coll. Scientific Papers*, **1**, p. 392, and **2**, pp. 57 and 345. See also Boltzmann, *Wien. Sitzungsber.* **66** [2] (1872), p. 324, 78 (1878), p. 733, 86 (1882), p. 63, and 88 (1883), p. 835. Also *Vorlesungen über Gastheorie*, **1**, p. 96.

or, in terms of the molecular-velocities,*

$$\mathfrak{D}_{12} = \frac{1}{3\pi\nu S_{12}^2} \sqrt{\bar{c}_1^2 + \bar{c}_2^2}. \qquad \ldots\ldots(260)$$

If the two kinds of molecules are of equal mass and size, this becomes

$$\mathfrak{D} = \frac{2}{3} \frac{\bar{c}}{\sqrt{2\pi\nu\sigma^2}}, \qquad \ldots\ldots(261)$$

which may be contrasted with Meyer's uncorrected formula (245).

Using Chapman's corrected formula (210) for the coefficient of viscosity,

$$\eta = 0\cdot 499 \frac{\rho\bar{c}}{\sqrt{2\pi\nu\sigma^2}},$$

equation (261) assumes the form

$$\mathfrak{D} = 1\cdot 336 \frac{\eta}{\rho}, \qquad \ldots\ldots(262)$$

agreeing almost exactly with equation (250) which was obtained on correcting Meyer's formula for persistence of velocities.

When the molecules of the two gases are unequal in mass and size, it is more difficult to compare equation (259) with equation (255) which was obtained by correcting Meyer's formula for persistence of velocities; there is the outstanding difference be-

---

* Meyer, using the value $\mathfrak{D}_{12}$ already explained (see footnote to p. 201), obtains a value for $\mathfrak{D}_{12}$ on Maxwell's theory equal to $\frac{3}{8}\pi$ times this, namely

$$\mathfrak{D}_{12} = \frac{1}{8\nu S_{12}^2} \sqrt{\bar{c}_1^2 + \bar{c}_2^2},$$

and this same value is given by Maxwell (*l.c. ante* and *Nature*, 8 (1873), p. 298). On the other hand, Stefan (*Wien. Sitzungsber.* 68 (1872), p. 323), Langevin (*Ann. de Chim. et de Phys.* [8], 5 (1905), p. 245), and Chapman in his first paper (*Phil. Trans.* 211 A (1912), p. 449) all arrived at the formula

$$\mathfrak{D}_{12} = \frac{3}{32\nu S_{12}^2} \sqrt{\bar{c}_1^2 + \bar{c}_2^2},$$

which is only three-quarters of the value above. Chapman and Langevin both extended their method to the general law of force μr^{-s}; their method was somewhat similar to that of Maxwell as given in Chap. IX (§ 201), but they assumed Maxwell's law of distribution to hold, so that their results were only exact for the case of $s = 5$, for which their result agrees with Maxwell's formula (264).

tween the two, that Meyer's formula depends on the ratio ν_1/ν_2, whereas Maxwell's formula (259) does not. In the limiting case of $\nu_1 = 0$, Meyer's formula reduces to formula (256), which is identical with Maxwell's formula (259) divided by the factor $(1 + \alpha_{12})$, where α_{12} is the small number defined in § 128. Thus the formulae approximate closely; they naturally cannot agree completely, since the one predicts slight variation with ν_1/ν_2, while the other predicts none at all.

Exact General Formulae

172. We have so far only obtained approximate formulae, and this only for the very special case of molecules which may be treated as elastic spheres.

When the molecules are treated as point centres of force repelling according to the law μr^{-s}, the methods of Chapman and Enskog, which are explained in Chap. IX below, allow the value of $\mathfrak{D}_{12}$ to be calculated, by successive approximations, to any desired degree of accuracy.

A first approximation, arrived at by the assumption that the components of velocity u, v, w relative to the velocity of mass-motion were distributed according to Maxwell's law, had been given by Langevin* in 1905; the same formula was given independently by Chapman† in 1911. To a first approximation the value of $\mathfrak{D}_{12}$ is found to be given by

$$\mathfrak{D}_{12} = \frac{3[\pi(m_1 + m_2)/hm_1 m_2]^{\frac{1}{2}}}{8(\nu_1 + \nu_2)\,[h\mu]^{\frac{2}{s-1}}\,I_1(s)\,\Gamma\!\left(3 - \dfrac{2}{s-1}\right)}, \qquad \ldots\ldots(263)$$

where, in the notation of § 195 below,

$$I_1(s) = 4\pi \int_{\alpha=0}^{\alpha=\infty} \cos^2 \tfrac{1}{2}\theta'\,\alpha\,d\alpha.$$

When $s = 5$, this is exact, and reduces to

$$\mathfrak{D}_{12} = \frac{1}{2h(m_1 m_2)^{\frac{1}{2}}(\nu_1 + \nu_2)\,A_1} \sqrt{\frac{m_1 + m_2}{\mu}}, \qquad \ldots\ldots(264)$$

where $\qquad\qquad A_1 = I_1(5) = 2 \cdot 6595,$

* *Ann. de Chim. et de Phys.* [8], **5** (1905), p. 245.
† *Phil. Trans.* **211** A (1912), p. 433.

a formula which had been derived by Maxwell as far back as 1866, by methods explained below (§ 201).

When s has a value other than 5, the exact value of $\mathfrak{D}_{12}$ will be obtained on multiplying the above approximate value of $\mathfrak{D}_{12}$ by a numerical multiplier.

Self-Diffusion

173. Chapman has obtained the following table of the values of this multiplier from calculations carried as far as a second approximation for special values of s in the case in which the two kinds of molecules are exactly similar (self-diffusion).

	$s = 5$	$s = 9$	$s = 17$	$s = \infty$
Multiplier =	1·000	1·004	1·008	1·015

It appears that the multiplier increases steadily as s increases from 5 upwards, and reaches its maximum of 1·015 when $s = \infty$ (elastic spheres). In the special case of elastic spheres, Chapman supposes that the multiplying factor, to a third approximation, would have the value 1·017, and that this is the value to which successive approximations are converging, to an accuracy of one part in a thousand. Assuming this, Chapman finds that the accurate value of the coefficient of self-diffusion for elastic spheres[*] is

$$\mathfrak{D} = \frac{0 \cdot 1520}{4 v \sigma^2 (h m)^{\frac{1}{2}}} \qquad \ldots \ldots (265)$$

$$= 1 \cdot 200 \frac{\eta}{\rho}. \qquad \ldots \ldots (266)$$

Pidduck,[†] following a method originated by Hilbert,[‡] based on the transformation of Boltzmann's characteristic equation (311), had previously arrived at the formula (265) for $\mathfrak{D}$, except that the number in the numerator, calculated to three places of decimals only, was given as 0·151.

---

[*] *Phil. Trans.* **217** A (1916), p. 172.
[†] *Proc. Lond. Math. Soc.* **15** (1915), p. 89.
[‡] *Math. Ann.* **72** (1912), p. 562.

Diffusion in a Mixture

174. Both Chapman* and Enskog† have shewn how to obtain, by successive approximations, formulae for the general value of $\mathfrak{D}_{12}$ in a mixture of gases. Chapman starts by taking the value given by equation (263) as a first approximation. Denoting this value by $(\mathfrak{D}_{12})_0$, the general value of $\mathfrak{D}_{12}$ is put in the form

$$\mathfrak{D}_{12} = \frac{(\mathfrak{D}_{12})_0}{1 - \epsilon_0}, \qquad \ldots\ldots(267)$$

where ϵ_0 is a small quantity, to be evaluated by successive approximations. As we have already seen, the value of ϵ_0 is zero for Maxwellian molecules repelling as the inverse fifth-power of the distance. In other cases the formulae, even when carried only to a second approximation, are extremely complicated, and the reader who wishes to study them in detail is referred to the original memoirs.‡

The principal interest of these general formulae lies in the amount of dependence of $\mathfrak{D}_{12}$ on the proportion of the mixture which is predicted by them. We have already seen that the approximate Maxwell-Stefan theory (§ 171) predicted that $\mathfrak{D}_{12}$ would be independent of this proportion; on the other hand, the theory of Meyer predicted very great dependence when the molecules of the two kinds of gas were of very unequal mass, although this, it is true, was greatly reduced when "persistence of velocities" was taken into account.

For the ratio of $\mathfrak{D}_{12}$ in the two extreme cases of $\nu_1/\nu_2 = 0$ and $\nu_1/\nu_2 = \infty$, Enskog§ gives the formula (accurate to a second approximation) for the case in which the molecules may be treated as elastic spheres,

$$\frac{\mathfrak{D}_{\nu_1=0}}{\mathfrak{D}_{\nu_2=0}} = \frac{1 + \dfrac{m_2^2}{12m_2^2 + 16m_1m_2 + 30m_1^2}}{1 + \dfrac{m_1^2}{12m_1^2 + 16m_1m_2 + 30m_2^2}}.$$

This may be compared with the value m_2/m_1 predicted by

---

* *Phil. Trans.* **217** A (1916), p. 166.

† *L.c. ante* (see footnote to p. 182).

‡ Chapman, *l.c.* equations (13·07), (13·28) ff.; Enskog, *l.c.* equations (168) ff.   § *L.c.* p. 103.

Meyer's uncorrected formula (§ 167), and with our formula (§ 170) obtained by correcting Meyer's formula for persistence of velocities. To take a definite instance, it will be found that when $m_1/m_2 = 10$, the predicted values are as follows:

Meyer	10·000
Meyer (corrected for persistence)	1·324
Chapman-Enskog	1·072

COMPARISON WITH EXPERIMENT

175. It will be convenient to consider first the experimental evidence as to how far the coefficient of diffusion depends on the proportion in which the gases are mixed. A series of experiments* have been made at Halle to test this question, a summary of which will be found in a paper by Lonius.†

Experiments were made on the pairs of gases H_2—O_2, H_2—N_2 and N_2—O_2 by Jackmann, on H_2—O_2 and H_2—CO_2 by Deutsch, and on He—A by Schmidt and Lonius. On every theory which has been considered, the greatest variation of $\mathfrak{D}_{12}$ with ν_1/ν_2 ought to occur when the ratio of the masses of the molecules differs most from unity. The following table‡ gives the values obtained for $\mathfrak{D}_{12}$ with different values of ν_1/ν_2 for the two pairs of gases for which this inequality of masses is greatest.

Pair of Gases (1, 2 respectively)	$\dfrac{\nu_1}{\nu_2}$	$\mathfrak{D}_{12}$ (observed)	Observer	$\mathfrak{D}_{12}$ (calculated) (Chapman)
H_2—CO_2	3	0·21351	Deutsch	0·212
	1	0·21774	,,	0·222
	$\frac{1}{3}$	0·22772	,,	0·226
He—A	2·65	0·24418	Lonius	0·248
	2·26	0·24965	,,	0·250
	1·66	0·25040	Schmidt	0·251
	1	0·25405	,,	0·254
	0·477	0·25626	Lonius	0·257
	0·311	0·26312	,,	0·259

* R. Schmidt, *Ann. d. Phys.* **14** (1904), p. 801, and the following Inaug. Dissertations: R. Schmidt (1904), O. Jackmann (1906), R. Deutsch (1907), and Lonius (1909).

† *Ann. d. Phys.* **29** (1909), p. 664. See also Chapman, *Phil. Trans.* **211** A (1912), p. 478.   ‡ Lonius, *l.c.* p. 676.

The last column gives the values calculated by Chapman from his theoretical formulae. In these calculations an absolute value of $\mathfrak{D}_{12}$ is assumed such as to make the mean of the calculated values of $\mathfrak{D}_{12}$ for each pair of gases equal to the mean of the observed values.

The degree of variation in $\mathfrak{D}_{12}$ predicted by the formula of Chapman is at least of the same order of magnitude as that actually observed. The superiority of Chapman's formulae over the two others already discussed is shewn in the following table, which gives a comparison between the extreme values of $\mathfrak{D}_{12}$ for He—A observed, and those predicted by these various formulae. (All values of $\mathfrak{D}_{12}$ are multiplied by a factor chosen so as to make $\mathfrak{D}_{12} = 1$ when $\nu_1 = \nu_2$.)

ν_1/ν_2	$\mathfrak{D}_{12}$ (observed)	$\mathfrak{D}_{12}$ (calculated)		
		(Chapman)	(Meyer; corrected)	(Meyer)
2·65	0·961	0·976	0·910	0·548
1·00	1·000	1·000	1·000	1·000
0·311	1·036	1·021	1·110	1·526

The foregoing discussion will have shewn that the actual variation of $\mathfrak{D}_{12}$ with the proportion of the mixture is, in any case, very slight. Consequently, throughout the remainder of this chapter we shall be content to disregard the dependence of $\mathfrak{D}_{12}$ on the ratio ν_1/ν_2.

Coefficient of Self-Diffusion

176. The formula which it is easiest to test numerically is that for self-diffusion, but the coefficient of self-diffusion of a gas into itself is not a quantity which admits of direct experimental determination.

A convenient plan, adopted by Lord Kelvin,* is to take a set of three gases for which the coefficients $\mathfrak{D}_{12}, \mathfrak{D}_{23}, \mathfrak{D}_{31}$ are known. All the quantities in formula (244) are then known with great accuracy except only S_{12}. Hence from the three values of

* *Baltimore Lectures*, p. 295.

$\mathfrak{D}_{12}, \mathfrak{D}_{23}, \mathfrak{D}_{31}$ we can calculate S_{12}, S_{23}, S_{31} and so deduce values of $\sigma_1, \sigma_2, \sigma_3$. Instead of comparing these values with other determinations of σ_1, σ_2 and σ_3, Lord Kelvin inserted them into formula (245) and so obtained the coefficients of self-diffusion of the three gases in question.

Lord Kelvin gives the following values of coefficients of inter-diffusivity of four gases, calculated from the experimental determinations of Loschmidt.

Gases

H_2 —(1)
O_2 —(2)
CO —(3)
CO_2—(4)

Pairs of gases			$\mathfrak{D}_{11}$	Pairs of gases			$\mathfrak{D}_{22}$
(12,	13,	23)	......1·32	(12,	13,	23)	......0·193
(12,	14,	24)	......1·35	(12,	14,	24)	......0·190
(13,	14,	34)	......1·26	(23,	24,	34)	......0·183
		Mean	$\overline{1·31}$			Mean	$\overline{0·189}$

Pairs of gases			$\mathfrak{D}_{33}$	Pairs of gases			$\mathfrak{D}_{44}$
(12,	13,	23)	......0·169	(12,	14,	24)	......0·106
(13,	14,	34)	......0·175	(13,	14,	34)	......0·111
(23,	24,	34)	......0·178	(23,	24,	34)	......0·109
		Mean	$\overline{0·174}$			Mean	$\overline{0·109}$

The agreement *inter se* of the values obtained by different sets of three gases gives a striking confirmation of the theory, except of course as regards the numerical multiplier which does not affect the values obtained for $\mathfrak{D}_{11}$, $\mathfrak{D}_{22}$, etc.

Somewhat similar although not entirely identical results were obtained by Boardman and Wild,[*] particularly experimenting with pairs of gases in which the molecules were nearly similar, so that the coefficient of interdiffusion of a pair of gases could be identified with the coefficient of self-diffusion of either. They obtained the following coefficients of self-diffusion, all at 15° C.:

Hydrogen	1·43
Nitrogen	0·203
Carbon-monoxide	0·211
Nitrous oxide	0·107
Carbon-dioxide	0·121

but their experiments did not yield completely consistent results,

[*] *Proc. Roy. Soc.* **162** A (1937), p. 511,

177. It remains to test the numerical multiplier. The calculations of Chapman, Enskog and Pidduck combine in predicting the relation (approximately)

$$\mathfrak{D} = 1 \cdot 200 \frac{\eta}{\rho}$$

for elastic spheres, while Maxwell's theory given in Chap. IX below predicts the relation (exactly)

$$\mathfrak{D} = 1 \cdot 543 \frac{\eta}{\rho}$$

for molecules repelling according to the inverse fifth-power of the distance.

In the following table, the second column gives the value assumed for η in the table of p. 183, the third column gives ρ, the fourth gives the value of $\mathfrak{D}$ calculated from Loschmidt's experiments, and the fifth gives the value of $\mathfrak{D}\rho/\eta$.

Gas	η (p. 183)	ρ	$\mathfrak{D}$ (p. 215)	$\dfrac{\mathfrak{D}\rho}{\eta}$
Hydrogen	0·0000857	0·0000899	1·31	1·37
Oxygen	0·0001926	0·001429	1·189	1·40
Carbon-monoxide	0·0001665	0·001250	0·174	1·30
Carbon-dioxide	0·000137	0·001977	0·109	1·58

Within the probable error of the experiments, it appears that $\mathfrak{D}\rho/\eta$ has values intermediate between the two values 1·200 and 1·543 predicted by theory for elastic spheres and inverse fifth-power molecules. Not only is this so, but the values of $\mathfrak{D}\rho/\eta$ vary between these limits in a manner which accords well with the knowledge we already have as to the laws of force (μr^{-s}) in the different gases concerned, as the following figures shew:

	Value of s (p. 174)	$\dfrac{\mathfrak{D}\rho}{\eta}$
Theory	∞	1·200
Hydrogen	11·3	1·37
Carbon-monoxide	8·7	1·30
Oxygen	7·4	1·40
Carbon-dioxide	5·6	1·58
Theory	5·0	1·543

It is somewhat remarkable that the values of $\mathfrak{D}\rho/\eta$ for hydrogen and carbon-monoxide, in which the molecules are comparatively "hard" (in the sense of § 147), approximate so closely to the value 1·34 predicted both by Meyer's corrected theory (§ 169) and by the Stefan-Maxwell theory (§ 171). This suggests that for ordinary natural molecules the two simpler theories may represent the processes at work with as great accuracy as the more elaborate theories of Chapman and Enskog so long as we remain in ignorance of the exact molecular structure which ought to be assumed in the latter.

Coefficient of Diffusion for Elastic Spheres

178. The following table gives the observed values of $\mathfrak{D}_{12}$ for a number of pairs of gases in which the molecules are comparatively hard, having values of s greater than 7·4 in the table of p. 174. The table gives also the values of S_{12} calculated from them by formula (260) (using values of $\bar{c}$ given on p. 56), and, in the last column, the values of S_{12} calculated from the coefficient of viscosity as on p. 183.

Gases	$\mathfrak{D}_{12}$ (observed)	S_{12} (calc. from $\mathfrak{D}_{12}$)	S_{12} (calc. from viscosity)
Hydrogen—Air	0·661	$3·23 \times 10^{-8}$	$3·23 \times 10^{-8}$
,, —Oxygen	0·679	3·18	3·17
Oxygen—Air	0·1775	3·69	3·68
,, —Nitrogen	0·174	3·74	3·70
Carbon-monoxide—Hydrogen	0·642	3·28	3·25
,, —Oxygen	0·183	3·65	3·70

The agreement between the two sets of values of S_{12} is as good as could reasonably be expected, providing a corresponding confirmation of formula (260). When one or both of the two kinds of molecules involved is softer than those in the foregoing table the agreement is still good, although less striking than that found above, as is shewn in the table below.

Gases	$\mathfrak{D}_{12}$ (observed)	S_{12} (calc. from $\mathfrak{D}_{12}$)	S_{12} (calc. from viscosity)
Carbon-dioxide—Hydrogen	0·538	$3·56 \times 10^{-8}$	$3·69 \times 10^{-8}$
„ —Air	0·138	4·03	4·20
„ —Carbon-monoxide	0·136	4·09	4·22
Nitrous oxide—Hydrogen	0·535	3·57	3·69
„ —Carbon-dioxide	0·0983	4·53	4·66
Ethylene—Hydrogen	0·486	3·75	4·14
„ —Carbon-monoxide	0·101	4·99	4·68

Thermal and Pressure Diffusion

179. In 1911 Enskog* predicted on theoretical grounds that diffusion must occur in any gas in which the temperature is not uniform throughout. The same is true when there is a pressure gradient, although this "pressure-diffusion" is of less practical importance than the "thermal diffusion" just mentioned. Chapman† predicted the same phenomena independently in 1916, and Dootson‡ confirmed the existence of thermal diffusion in 1917. The discussion of these phenomena is postponed to the end of the next chapter.

RANDOM MOTIONS AND BROWNIAN MOVEMENTS

180. In §154 we obtained the equation of conduction of heat in a gas in the form

$$\frac{dT}{dt} = \frac{\eta}{\rho} \frac{\partial^2 T}{\partial z^2}, \qquad \dots\dots(268)$$

while in §165 we obtained as the equation of diffusion (equation (243))

$$\frac{dv_1}{dt} = \mathfrak{D}_{12} \frac{\partial^2 v_1}{\partial z^2}. \qquad \dots\dots(269)$$

Replacing $\mathfrak{D}_{12}$ by the value η/ρ appropriate for self-diffusion, this becomes

$$\frac{dv_1}{dt} = \frac{\eta}{\rho} \frac{\partial^2 v_1}{\partial z^2}. \qquad \dots\dots(270)$$

* *Phys. Zeitschr.* **12** (1911), p. 533.
† *Phil. Trans.* **217** A (1916), p. 164.   ‡ *Phil. Mag.* **33** (1917), p. 248.

We notice at once that this is of the same form as the equation of conduction of heat (268), so that heat is diffused through a gas at the same rate and according to the same laws as any selected group of molecules. There is, of course, a reason for this. If there were no collisions, both heat and molecules would proceed through the gas at the same speed, the speed of molecular motion; when collisions impede progress, they affect the progress of both heat and molecules in the same way.

It will be understood that here, as throughout the remainder of this chapter, we are studying the phenomena in their broad outlines only, and are not striving after numerical exactitude. The true value of $\mathfrak{D}_{12}$ differs from the value η/ρ we have assigned to it by a numerical factor, but we do not concern ourselves with this in the present discussion.

The simplest solution of equation (269) is

$$v_1 = \frac{A}{\sqrt{t}} e^{-\frac{z^2}{4\mathfrak{D}t}}, \qquad \ldots\ldots(271)$$

where A is any constant. This is the solution for a group of molecules which start from the plane $z = 0$ at the instant $t = 0$, and gradually diffuse through the gas. There are, of course, precisely similar solutions to equations (268) and (270). We shall now see how the same equations and solutions could have been obtained directly from a detailed study of the molecular movements in the gas.

Random Motions

181. Let us consider the random motion of a single molecule as it is hit about by collisions with other molecules. As we are concerned with general principles rather than numerical exactitude, let us simplify the problem to the extreme limit by supposing that all free paths are of the same length l, and that all are described parallel to the axis of z; we shall further suppose that for each free path the directions of z increasing and of z decreasing are equally likely.

Any molecule which has described a number N of free paths will have advanced along the axis of z by a distance

$$\pm l \pm l \pm l \pm \ldots \text{ (to } N \text{ terms).}$$

Each of the signs in this series may be either + or −, so that there are 2^N possibilities in all. For some arrangements of signs, the molecule will have advanced a distance sl along the axis of z. For this to be the case, there must be p positive and q negative signs, where p, q are given by

$$p+q = N, \quad p-q = s. \qquad \ldots\ldots(272)$$

The number of ways in which + and − signs can be arranged to give this result is of course

$$\frac{N!}{p!\,q!},$$

so that the chance P that the molecule shall have advanced a distance sl after N free paths is

$$P = \frac{N!}{p!\,q!\,2^N}. \qquad \ldots\ldots(273)$$

This provides a solution to what is commonly known as the problem of the "random walk". A man takes a walk of N steps of a yard each, and each step is as likely to be forward as backward; what is the chance that the end of his walk will find him s yards in front of his starting point? Obviously the answer is given by formula (273).

182. In the applications of this formula to kinetic theory problems, N, p and q are all large numbers, so that the value of P can be simplified by using Stirling's approximation

$$n! = \sqrt{2\pi n}\left(\frac{n}{e}\right)^n,$$

to the value of $n!$ when n is large. Using this relation, equation (273) takes the form

$$P = \frac{1}{\sqrt{\pi}}\frac{(\tfrac{1}{2}N)^{N+\frac{1}{2}}}{p^{p+\frac{1}{2}}q^{q+\frac{1}{2}}}.$$

From equations (272) we find

$$p = \tfrac{1}{2}(N+s) = \tfrac{1}{2}N\left(1+\frac{s}{N}\right),$$

$$q = \tfrac{1}{2}(N-s) = \tfrac{1}{2}N\left(1-\frac{s}{N}\right),$$

so that $\qquad P = \dfrac{1}{\sqrt{\pi}} \dfrac{1}{(\frac{1}{2}N)^{\frac{1}{2}} \left(1 + \dfrac{s}{N}\right)^{p+\frac{1}{2}} \left(1 - \dfrac{s}{N}\right)^{q+\frac{1}{2}}}. \qquad \ldots\ldots(274)$

When N is made very large, the forward and backward paths must nearly cancel one another, so that in general s will be small compared with N, and $p + \frac{1}{2}$, $q + \frac{1}{2}$ may each be put equal to $\dfrac{1}{2N}$. Thus we may write

$$\left(1 + \frac{s}{N}\right)^{p+\frac{1}{2}} \left(1 - \frac{s}{N}\right)^{q+\frac{1}{2}} = \left(1 - \frac{s^2}{N^2}\right)^{\frac{1}{2}N} = e^{-\frac{s^2}{2N}},$$

and P, as given by formula (274), assumes the approximate value

$$P = \sqrt{\frac{2}{\pi N}} e^{-\frac{s^2}{2N}}. \qquad \ldots\ldots(275)$$

This is the chance that after describing N free paths each of length l, the molecule shall have advanced a distance sl.

183. Let us now replace sl by z and N by ct/l, where c is the speed of molecular motion and t the time the motion has progressed. Formula (275) becomes

$$P = \sqrt{\frac{2l}{\pi ct}} e^{-\frac{z^2}{2lct}}, \qquad \ldots\ldots(276)$$

which is in all essentials identical with the equation (271) already found. Equation (271) gives the density of molecules at a distance z from the plane $z = z_0$, while equation (276) gives the probability that any single molecule shall be found at this distance from $z = z_0$. Clearly the values of P and ν_1 must be the same except for a multiplying constant.

For the exponentials to be the same, we must have

$$\mathfrak{D} = \tfrac{1}{2}lc = \frac{\eta}{\rho}, \qquad \ldots\ldots(277)$$

where η has the uncorrected value obtained in equation (47). Thus the analysis just given would have enabled us to calculate $\mathfrak{D}$ accurately except for the numerical adjustments required by variations in the lengths of free path, persistence of velocities, and so forth. Its true importance is, however, that it presents us with a vivid picture of the process of diffusion; we now see diffusion as a random walk.

When P is given by formula (276), the graph of P as a function of z is an exponential curve like the thin curve in fig. 21 (p. 118). The maximum ordinate is always at the origin—no matter how long diffusion goes on for, initial condensations are never entirely smoothed out.

Using the formula of Appendix VI (p. 306), we readily find that the average numerical value of z is $(2lct/\pi)^{\frac{1}{2}}$ or $0\cdot798\,\sqrt{lct}$. Thus as diffusion proceeds, the average, or any other proportionate, value of s increases only as $\sqrt{t}$; travelling at random, we must take four times as many steps to travel two miles as to travel one.

Brownian Movements

184. The foregoing ideas are applicable to the Brownian movements, although in a somewhat modified form. Each Brownian particle will perform a sort of random walk, but as its mass is much greater than that of the particles it encounters, persistence of velocities becomes all important.

Following Einstein[*] and Smoluchowski,[†] we may treat all collisions with other molecules as forming a statistical group, and regard the Brownian particle as ploughing its way through a viscous fluid. We accordingly suppose the particle to experience a viscous drag in the direction opposite to its motion. We may suppose this viscous force to have components

$$-\epsilon\eta\,\frac{dx}{dt}, \quad -\epsilon\eta\,\frac{dy}{dt}, \quad -\epsilon\eta\,\frac{dz}{dt},$$

where η is the coefficient of viscosity and ϵ is a second coefficient which depends on the size and shape of the particle, as well as on the physical properties both of the particle and of the fluid through which it moves. The equations of motion of the particle are now

$$X-\epsilon\eta\,\frac{dx}{dt} = m\,\frac{d^2x}{dt^2}, \text{ etc.,} \qquad \ldots\ldots(278)$$

---

[*] *Ann. d. Physik*, **17** (1905), p. 549 and **19** (1906), p. 371.
[†] *Bulletin Acad. Sci. Cracovie* (1906), p. 577.

where X, Y, Z are the components of any extraneous forces which may act on the particle.

Exactly as with the virial equation of Clausius, we multiply these three equations of motion by x, y, z, and add corresponding sides. This gives, after slight transformation,

$$xX + yY + zZ - \tfrac{1}{2}\epsilon\eta \frac{d}{dt}(x^2+y^2+z^2) = \tfrac{1}{2}m \frac{d^2}{dt^2}(x^2+y^2+z^2) - mc^2.$$

Integrating through an interval of time t, and averaging over a number of molecules, we obtain

$$-\tfrac{1}{2}\epsilon\eta\Delta(x^2+y^2+z^2) = -\overline{mc^2}t = -3RTt,$$

where $\Delta(x^2+y^2+z^2)$ denotes the change in $x^2+y^2+z^2$ in time t. Thus, finally,

$$\Delta(x^2+y^2+z^2) = \frac{6RT}{\epsilon\eta}t. \qquad \ldots\ldots(279)$$

Again, as in the random walk, the distance travelled in a time t is proportional to $\sqrt{t}$. It is also, naturally enough, proportional to $\sqrt{T}$, or to $\bar{c}$. Both these relations have been confirmed experimentally by Perrin and many others.

If the Brownian particles are spheres of radius a, Stokes's law gives the value of ϵ as $6\pi a$, and equation (279) assumes the form

$$\Delta(x^2+y^2+z^2) = \frac{RT}{\pi a\eta}t. \qquad \ldots\ldots(280)$$

This equation has been used as a means of determining R, and hence Loschmidt's and Avogadro's numbers, with satisfactory results (cf. § 16, above).

185. So far we have discussed only the average value of $x^2+y^2+z^2$ for all the particles. The distribution of the individual particles can be discussed in a similar manner, but is best found by treating the motion as one of diffusion. As in § 180, the distribution of density must satisfy an equation of the form

$$\frac{d\nu}{dt} = K\left(\frac{\partial^2\nu}{\partial x^2}+\frac{\partial^2\nu}{\partial y^2}+\frac{\partial^2\nu}{\partial z^2}\right), \qquad \ldots\ldots(281)$$

where K is a constant, which is of the nature of a coefficient of diffusion.

Analogous to the solution (271) of our former equation, we obtain as a solution of equation (281),

$$\nu = \frac{A}{t^{\frac{3}{2}}} e^{-\frac{x^2+y^2+z^2}{4Kt}}, \qquad \ldots\ldots(282)$$

where A is a constant. For such a distribution of particles, we readily find that the average value of $x^2+y^2+z^2$ is $6Kt$. The solution (282) is appropriate to a group of particles starting at the origin at time $t = 0$, and gradually diffusing through space; for such a group we have already found the average value of $x^2+y^2+z^2$ to be $\dfrac{6RT}{\epsilon\eta} t$ (equation (279)). Equating this to $6Kt$, we find

$$K = \frac{RT}{\epsilon\eta}.$$

This, then, is the value of K in the equation of diffusion (281), and with this value of K formula (282) will give the distribution in space of a group of particles which started at the origin at the instant $t = 0$. On assigning this value to K in the more general equation (281), we can trace the motion resulting from any distribution of Brownian particles.

Chapter IX

GENERAL THEORY OF A GAS NOT IN A STEADY STATE

186. The present chapter will deal with a variety of problems in which the gas is not in a steady state, so that the velocities of the molecules are not distributed according to Maxwell's law. In particular we shall discuss the analogy between a gas and the fluid of hydrodynamical theory, and shall examine new methods of dealing with the viscosity, heat conduction and diffusion of a gas. The distribution of velocities will be supposed to be given by the quite general law

$$f(u, v, w, x, y, z),$$

where the symbols have the same meaning as elsewhere in the book, particularly in § 22.

Hydrodynamical Equation of Continuity

187. We first fix our attention on a small element of volume $dx\,dy\,dz$ inside the gas, having its centre at ξ, η, ζ and bounded by the six planes parallel to the coordinate planes

$$x = \xi \pm \tfrac{1}{2}dx, \quad y = \eta \pm \tfrac{1}{2}dy, \quad z = \zeta \pm \tfrac{1}{2}dz.$$

The number of molecules of class A (defined as on p. 105) which cross the plane $x = \xi - \tfrac{1}{2}dx$ into this element of volume in time dt will be

$$\nu f(u, v, w, \xi - \tfrac{1}{2}dx, \eta, \zeta)\,dy\,dz\,du\,dv\,dw\,u\,dt.$$

Similarly the number of molecules of class A which cross the plane $x = \xi + \tfrac{1}{2}dx$ out of the element of volume is given by

$$\nu f(u, v, w, \xi + \tfrac{1}{2}dx, \eta, \zeta)\,dy\,dz\,du\,dv\,dw\,u\,dt.$$

By subtraction, we find that the element experiences a net loss of molecules of class A, caused by motion through the two faces perpendicular to the axis of x, of amount

$$\frac{\partial}{\partial x}[\nu f(u,\ v,\ w)]\,dx\,dy\,dz\,du\,dv\,dw\,u\,dt,$$

in which the differential coefficient is evaluated at ξ, η, ζ. The net loss of molecules of class A caused by motion through all six faces is therefore

$$\left(u\frac{\partial}{\partial x}+v\frac{\partial}{\partial y}+w\frac{\partial}{\partial z}\right)[\nu f(u,\,v,\,w)]\,du\,dv\,dw\,dx\,dy\,dz\,dt. \quad ...(283)$$

On integrating over all values of u, v and w, we obtain the total number of molecules which are lost to the element $dx\,dy\,dz$ in time dt. We may again write

$$\iiint uf(u,\,v,\,w)\,du\,dv\,dw = u_0, \text{ etc.,}$$

where u_0, v_0, w_0 are the components of mass-velocity of the gas at x, y, z; this number is now seen to be

$$\left(\frac{\partial}{\partial x}(\nu u_0)+\frac{\partial}{\partial y}(\nu v_0)+\frac{\partial}{\partial z}(\nu w_0)\right)dx\,dy\,dz\,dt. \quad ......(284)$$

The number of molecules in the element at time t is, however, $\nu\,dx\,dy\,dz$, and that at time $t+dt$ is $\left(\nu+\dfrac{d\nu}{dt}dt\right)dx\,dy\,dz$. Thus the net loss must be

$$-\frac{d\nu}{dt}dt\,dx\,dy\,dz.$$

Equating this to expression (284), we obtain

$$\frac{d\nu}{dt}+\frac{\partial}{\partial x}(\nu u_0)+\frac{\partial}{\partial y}(\nu v_0)+\frac{\partial}{\partial z}(\nu w_0)=0. \quad ......(285)$$

If we multiply throughout by m, the mass of a molecule, and replace νm by ρ, the density of the gas, this becomes

$$\frac{d\rho}{dt}+\frac{\partial}{\partial x}(\rho u_0)+\frac{\partial}{\partial y}(\rho v_0)+\frac{\partial}{\partial z}(\rho w_0)=0. \quad ......(286)$$

This is the hydrodynamical equation of continuity, expressing the permanence of the molecules of the gas—in other words, the conservation of mass.

The Hydrodynamical Equations of Motion

188. There is a somewhat similar equation expressing that momentum is conserved, or is changed only by the operation of external forces.

Each molecule of class A carries x-momentum of amount mu, so that the loss of a molecule of class A to the element $dx\,dy\,dz$ involves a loss of x-momentum equal to mu. The number of molecules of class A which are lost to the element in time dt is given by expression (283), so that the loss of x-momentum from this cause is mu times expression (283) and the total loss is

$$m\,dx\,dy\,dz\,dt \iiint \left(u^2\frac{\partial}{\partial x}+uv\frac{\partial}{\partial y}+uw\frac{\partial}{\partial z}\right)[vf(u,\,v,\,w)]\,du\,dv\,dw.$$

This may be written in the form

$$m\,dx\,dy\,dz\,dt\left(\frac{\partial}{\partial x}(v\overline{u^2})+\frac{\partial}{\partial y}(v\overline{uv})+\frac{\partial}{\partial z}(v\overline{uw})\right),\quad \ldots\ldots(287)$$

where

$$\overline{u^2}=\iiint u^2 f(u,\,v,\,w)\,du\,dv\,dw,$$

$$\overline{uv}=\iiint uv f(u,\,v,\,w)\,du\,dv\,dw,\ \text{etc.},$$

so that $\overline{u^2}$, $\overline{uv}$, etc. are the mean values of u^2, uv, etc. at the point x, y, z.

There is a further loss or gain of momentum from the action of forces on the molecules inside the element $dx\,dy\,dz$. A force X acting on a molecule in the direction of Ox causes a gain of x-momentum $X\,dt$ in a time dt. Combining the sum of these gains with the loss given by expression (287), we find for the net increase of momentum inside the element $dx\,dy\,dz$ in time dt,

$$\left[\Sigma X-m\,dx\,dy\,dz\left(\frac{\partial}{\partial x}(v\overline{u^2})+\frac{\partial}{\partial y}(v\overline{uv})+\frac{\partial}{\partial z}(v\overline{uw})\right)\right]dt,\quad \ldots\ldots(288)$$

where Σ denotes summation over all the molecules which were inside the element $dx\,dy\,dz$ at the beginning of the interval dt.

The total x-momentum inside the element $dx\,dy\,dz$ at time t was, however, $mvu_0\,dx\,dy\,dz$. The gain in time dt is accordingly

$$\frac{d}{dt}(vu_0)\,m\,dx\,dy\,dz\,dt,$$

and on equating this to expression (288), we obtain

$$\left[\frac{d}{dt}(vu_0)+\frac{\partial}{\partial x}(v\overline{u^2})+\frac{\partial}{\partial y}(v\overline{uv})+\frac{\partial}{\partial z}(v\overline{uw})\right]m\,dx\,dy\,dz=\Sigma X.$$

$$\ldots\ldots(289)$$

These and the similar equations in y, z are the equations of motion of the gas, expressing that momentum is changed only by the action of external forces.

189. In this equation, let us write

$$u = u_0 + \mathbf{U}, \text{ etc.,}$$

thus dividing the motion into the mass motion of the gas as a whole, and the thermal motions of the molecules. We have $\bar{u} = u_0$, and $\overline{\mathbf{U}} = 0$, so that $\overline{uv} = u_0 v_0 + \overline{\mathbf{UV}}$, and similarly for $\overline{u^2}$ and $\overline{uw}$.

If we multiply equation (285) throughout by u_0, and subtract corresponding sides from equation (289), the latter equation becomes

$$\nu \left(\frac{d}{dt} + u_0 \frac{\partial}{\partial x} + v_0 \frac{\partial}{\partial y} + w_0 \frac{\partial}{\partial z} \right) u_0 \, m \, dx \, dy \, dz$$

$$= \Sigma X - \left[\frac{\partial}{\partial x} (\nu \overline{\mathbf{U}^2}) + \frac{\partial}{\partial y} (\nu \overline{\mathbf{UV}}) + \frac{\partial}{\partial z} (\nu \overline{\mathbf{UW}}) \right] m \, dx \, dy \, dz.$$

$$\dots\dots(290)$$

We proceed to examine more closely the system of forces which act upon those molecules of which the centres are inside the element $dx \, dy \, dz$—the system of forces which we have so far been content to denote by ΣX, ΣY, ΣZ.

These forces will arise in part from the action of an external field of force upon the molecules of the gas, and in part from the actions of the molecules upon one another.

If there is a field of external force of components $\varXi$, H, Z per unit mass, the contribution to ΣX will be

$$\varXi m \nu \, dx \, dy \, dz. \qquad\qquad \dots\dots(291)$$

The remaining contribution of ΣX arises from the intermolecular forces. As regards the forces between a pair of molecules, both of which are inside the element $dx \, dy \, dz$, we see that, since action and reaction are equal and opposite, the total contribution to ΣX will be *nil*. We are left with forces between pairs of molecules such that one is inside and the other outside the element $dx \, dy \, dz$, that is to say, intermolecular forces which act across the boundary of this element.

Let the sum of the components of all such forces across a plane perpendicular to the axis of x be $\varpi_{xx}, \varpi_{xy}, \varpi_{xz}$ per unit area, and let us adopt a similar notation as regards forces acting across planes perpendicular to the axes of y and z. Then the contribution to ΣX from forces acting across the two planes $x = \xi \pm \frac{1}{2}dx$ is

$$(\varpi_{xx})_{x=\xi-\frac{1}{2}dx}\,dy\,dz - (\varpi_{xx})_{x=\xi+\frac{1}{2}dx}\,dy\,dz = -\frac{\partial \varpi_{xx}}{\partial x}dx\,dy\,dz.$$

On adding similar contributions from the other planes, we find that the x-component of all the molecular forces which act on the element $dx\,dy\,dz$ can be put in the form

$$-\left(\frac{\partial \varpi_{xx}}{\partial x} + \frac{\partial \varpi_{yx}}{\partial y} + \frac{\partial \varpi_{zx}}{\partial z}\right)dx\,dy\,dz.$$

Combining this contribution to ΣX with that already found in expression (291), we find, as the whole value of ΣX,

$$\Sigma X = \left[mv\Xi - \left(\frac{\partial \varpi_{xx}}{\partial x} + \frac{\partial \varpi_{yx}}{\partial y} + \frac{\partial \varpi_{zx}}{\partial z}\right)\right]dx\,dy\,dz,$$

and there are, of course, similar equations giving the values of ΣY and ΣZ.

Inserting this value for ΣX into equation (290), dividing throughout by $dx\,dy\,dz$ and replacing mv by ρ, we obtain

$$\rho\left(\frac{d}{dt} + u_0\frac{\partial}{\partial x} + v_0\frac{\partial}{\partial y} + w_0\frac{\partial}{\partial z}\right)u_0$$

$$= \rho\Xi - \frac{\partial}{\partial x}\left(\varpi_{xx} + \rho\overline{\mathsf{U}^2}\right) - \frac{\partial}{\partial y}\left(\varpi_{yx} + \rho\overline{\mathsf{UV}}\right) - \frac{\partial}{\partial z}\left(\varpi_{zx} + \rho\overline{\mathsf{UW}}\right).$$

This may be compared with the standard hydrodynamical equation,

$$\rho\left(\frac{d}{dt} + u_0\frac{\partial}{\partial x} + v_0\frac{\partial}{\partial y} + w_0\frac{\partial}{\partial z}\right)u_0 = \rho\Xi - \frac{\partial P_{xx}}{\partial x} - \frac{\partial P_{xy}}{\partial y} - \frac{\partial P_{xz}}{\partial z},$$

where u_0, v_0, w_0 is the velocity of the fluid at x, y, z, and P_{xx}, P_{xy}, P_{xz} are the components of stress per unit area on unit face perpendicular to Ox.

We see at once that the gas may be treated as a hydrodynamical fluid with stress components given by

$$\left.\begin{aligned} P_{xx} &= \varpi_{xx} + \rho\overline{\mathsf{U}^2}, \\ P_{yx} &= \varpi_{yx} + \rho\overline{\mathsf{UV}}, \text{ etc.} \end{aligned}\right\} \qquad \ldots\ldots(292)$$

Gas in a Steady State

190. In the special case in which the gas is at rest and in a steady state, the molecular velocities will be distributed according to Maxwell's law at every point, so that

$$\rho \overline{U^2} = \rho \overline{u^2} = \nu \overline{mu^2} = \frac{\nu}{2h},$$

as in § 88, and

$$\overline{UV} = \overline{VW} = \overline{WU} = 0.$$

Thus of the system of pressures specified by equations (292), that part which arises from molecular agitation reduces to a simple hydrostatical pressure of amount $\nu/2h$. Clearly the ϖ system of pressures, arising from the intermolecular forces, must also reduce to a simple hydrostatical pressure ϖ, and we therefore have

$$P_{xx} = P_{yy} = P_{zz} = \varpi + \frac{\nu}{2h},$$
$$P_{xy} = P_{xz} = P_{yz} = 0.$$

The total pressure at the point x, y, z is therefore a simple hydrostatical pressure of amount P given by

$$P = \varpi + \frac{\nu}{2h}, \qquad \qquad \text{......(293)}$$

an equation which may be compared with the virial equation (80).

Maxwell's Equations of Transfer

191. When we discussed the transfer of molecules across the boundary of a small volume $dx\,dy\,dz$ in § 187, we arrived at the hydrodynamical equation of continuity. When we discussed the transfer of momentum in § 188, we arrived at the hydrodynamical equations of motion. Let us now, following a procedure first developed by Maxwell,* consider the problem of the transfer of any quantity Q which depends solely on the velocity of a molecule, as for instance u^2 or uv.

* *Phil. Trans.* **157** (1886), p. 1, or *Collected Works*, **2**, p. 26.

Let the mean value of $\bar{Q}$ at any point x, y, z be denoted by $\bar{Q}$, so that

$$\bar{Q} = \iiint f(u, v, w)\, Q\, du\, dv\, dw.$$

The number of molecules inside the small element of $dx\, dy\, dz$ at this point is $\nu dx\, dy\, dz$, so that ΣQ, the aggregate amount of Q inside the element, is given by

$$\Sigma Q = \nu \bar{Q} dx\, dy\, dz. \qquad \ldots\ldots(294)$$

There are various causes of change in ΣQ. In the first place some molecules will leave the element $dx\, dy\, dz$, taking a certain amount of Q with them. As in § 187, the total number of molecules of class A lost to the element $dx\, dy\, dz$ in time dt is

$$dx\, dy\, dz\, dt \left(u\frac{\partial}{\partial x} + v\frac{\partial}{\partial y} + w\frac{\partial}{\partial z} \right)(\nu f)\, du\, dv\, dw,$$

so that the total amount of Q lost by motion into and out of the element is

$$dx\, dy\, dz\, dt \iiint \left[\left(u\frac{\partial}{\partial x} + v\frac{\partial}{\partial y} + w\frac{\partial}{\partial z} \right)(\nu f) \right] Q\, du\, dv\, dw.$$

If we write

$$\iiint u Q f(u, v, w)\, du\, dv\, dw = \overline{uQ}, \text{ etc.,}$$

so that $\overline{uQ}$ is the mean value of uQ averaged over all the molecules in the neighbourhood of the point x, y, z, this loss of Q can be put in the form

$$dx\, dy\, dz\, dt \left[\frac{\partial}{\partial x}(\nu \overline{uQ}) + \frac{\partial}{\partial y}(\nu \overline{vQ}) + \frac{\partial}{\partial z}(\nu \overline{wQ}) \right]. \qquad \ldots\ldots(295)$$

A second cause of change in ΣQ is the action of external forces on the molecules. For any single molecule, the rate of change in Q from this cause is given by

$$\frac{dQ}{dt} = \frac{\partial Q}{\partial u}\frac{du}{dt} + \frac{\partial Q}{\partial v}\frac{dv}{dt} + \frac{\partial Q}{\partial w}\frac{dw}{dt} = \frac{1}{m}\left(X\frac{\partial Q}{\partial u} + Y\frac{\partial Q}{\partial v} + Z\frac{\partial Q}{\partial w} \right),$$

where X, Y, Z are the components of the external force acting on the molecule. Thus in time dt the total value of ΣQ experiences an increase

$$dx\,dy\,dz\,dt\,\frac{\nu}{m}\left[X\left(\overline{\frac{\partial Q}{\partial u}}\right) + Y\left(\overline{\frac{\partial Q}{\partial v}}\right) + Z\left(\overline{\frac{\partial Q}{\partial w}}\right) \right], \qquad ......(296)$$

where again the bar over a quantity indicates an average taken over all molecules.

Lastly, ΣQ may be changed by collisions between molecules. If Q is any one of the quantities which have previously been denoted by $\chi_1, \chi_2, \ldots \chi_5$, namely, the mass, energy, and the three components of momentum of a molecule, there is no such change, but if Q is any other function of the velocities such changes will occur. In general let us denote the increase in ΣQ which is caused in the element $dx\,dy\,dz$ by collisions in time dt by

$$dx\,dy\,dz\,dt\,\Delta Q. \qquad ......(297)$$

Then the total change in the aggregate value of Q for all the molecules in $dx\,dy\,dz$ is given by the sum of expressions (295), (296) and (297). The aggregate value of Q is however given by expression (294). Equating the change in this to the sum of the three contributions just calculated, we obtain

$$\frac{d}{dt}(\nu\overline{Q}) = -\left[\frac{\partial}{\partial x}(\nu\overline{uQ}) + \frac{\partial}{\partial y}(\nu\overline{vQ}) + \frac{\partial}{\partial z}(\nu\overline{wQ}) \right]$$
$$+ \frac{\nu}{m}\left[X\left(\overline{\frac{\partial Q}{\partial u}}\right) + Y\left(\overline{\frac{\partial Q}{\partial v}}\right) + Z\left(\overline{\frac{\partial Q}{\partial w}}\right) \right] + \Delta Q. \qquad ......(298)$$

This is the general equation of transfer of Q. If we put $Q = m$, the mass of a molecule, the equation reduces to (286), the hydrodynamical equation of continuity, while if we put $Q = mu$, the x-momentum, the equation becomes identical with equation (289), the equation of motion of the gas.

If we multiply both sides of equation (285) by $\overline{Q}$ and subtract from corresponding sides of equation (298), this latter equation assumes the alternative form

$$\nu\frac{d\overline{Q}}{dt} = \Sigma\left[\overline{Q}\frac{\partial}{\partial x}(\nu u_0) - \frac{\partial}{\partial x}(\nu\overline{uQ}) + \frac{\nu}{m}X\left(\overline{\frac{\partial Q}{\partial u}}\right) \right] + \Delta Q, \qquad ......(299)$$

where Σ denotes summation with respect to x, y and z. This gives a new, and more useful, form of the equation of transfer.

The difficulties of the equation are centred in the calculation of ΔQ. Maxwell gave a general method for the calculation of ΔQ, which is explained in § 196 below, but found himself restricted by mathematical difficulties to the particular case in which the molecules attracted or repelled according to the law of force μr^{-5}.

On identifying Q with various functions of u, v and w, Maxwell obtained equations which were suitable for the discussion of the problems of viscosity, conduction of heat, and diffusion. The numerical results he calculated have already been quoted in their appropriate places.

In 1915 Chapman, in a paper of great power, succeeded in extending Maxwell's analysis, so as to apply to the general case of molecules repelling according to the law μr^{-s}. Chapman's analysis is unfortunately too long to be given here even in outline, but his main results have already been quoted and discussed. They include of course Maxwellian molecules ($s = 5$) and elastic spheres ($s = \infty$) as special cases.

Time of Relaxation

192. Maxwell's general equation of transfer (299) assumes its simplest form when the gas has no mass-motion, so that

$$u_0 = v_0 = w_0 = 0,$$

and when no external forces act, so that $X = Y = Z = 0$. If the distribution of velocities is the same throughout the gas, the equation takes the form

$$\nu \frac{d\overline{Q}}{dt} = \Delta Q,$$

expressing that the whole change in $\overline{Q}$ is caused by collisions.

Taking $Q = u^2 - v^2$, in this equation, Maxwell calculated that

$$\Delta Q = -\tau \nu^2 (\overline{u^2} - \overline{v^2}),$$

where τ is a constant depending only on the structure of the molecules. Similarly, on taking $Q = uv$, he calculated that

$$\Delta Q = -\tau \nu^2 \overline{uv},$$

where τ is the same constant as before. Thus it appears that $\overline{u^2 - v^2}$ and $\overline{uv}$ both satisfy an equation of the form

$$\frac{\partial \phi}{\partial t} = -\tau \nu \phi,$$

of which the solution,

$$\phi = \phi_0 e^{-\tau \nu t}, \qquad \ldots\ldots(300)$$

shews that ϕ decreases exponentially with the time at such a rate that it is reduced to $1/e$ times its original value in a time $1/\tau\nu$. Maxwell calls this the "time of relaxation".

This time of relaxation measures the rate at which deviations from Maxwell's law of distribution will subside. A glance at equations (292) shews that it must also measure the rate at which inequalities of pressure must subside, as also the shear stresses P_{yx}, P_{xy}, etc.

Maxwell shewed that τ is related to the coefficient of viscosity η by the relation

$$\tau = \frac{p}{\eta \nu},$$

where p is the hydrostatic pressure, and this provides the simplest means of evaluating τ. Replacing p by its usual value νRT, this relation becomes

$$\tau = \frac{RT}{\eta} . \qquad \ldots\ldots(301)$$

To take a definite instance, the value of η for air at $0°$ C. is $0 \cdot 000172$, and $RT = 3 \cdot 69 \times 10^{-14}$, so that $\tau = 2 \cdot 15 \times 10^{-10}$, and the time of relaxation is

$$\frac{1}{\tau\nu} = \frac{1}{6 \times 10^9} \text{ seconds.}$$

It is, as we should expect, comparable with the time of describing a free path. Thus by the time that a few free paths have been described by each molecule, the exponential on the right hand of equation (300) has already become very small, so that the gas is almost in the state specified by Maxwell's law.

The Lorentz-Enskog Analysis

193. In 1906, many years after the publication of Maxwell's work, an investigation given by Lorentz* for a different purpose suggested a new line of attack on Maxwell's problem. This was developed and carried to a successful termination by Enskog† in 1917.

When a gas is not in equilibrium or in a steady state, the distribution of velocities will no longer be given by Maxwell's law. If, however, the law of distribution f is known at the instant t, it is clearly possible in principle to follow out the motion of each group of molecules, calculate the changes in f, and so obtain the law of distribution at the next instant $t+dt$, and similarly at every subsequent instant. Thus the law of distribution is determined for all time when its value is given at any one instant. The function f must, then, satisfy an equation giving df/dt in terms of f. We proceed to investigate the general form of this equation, following a method given by Boltzmann.‡

194. Let the molecules be supposed to move in a permanent field of force, such that a molecule at x, y, z is acted on by a force (X, Y, Z) per unit mass. Then the equations of motion of a molecule, apart from collisions, are

$$\frac{du}{dt} = X, \quad \frac{dv}{dt} = Y, \quad \frac{dw}{dt} = Z.$$

The number of molecules which at any instant t have velocity components u, v, w within a small range $du\,dv\,dw$, and coordinates x, y, z within a small range $dx\,dy\,dz$, will be taken to be

$$\nu f(u, v, w, x, y, z, t)\,du\,dv\,dw\,dx\,dy\,dz. \qquad \ldots\ldots(302)$$

Let these molecules pursue their natural motion for a time dt. At the end of this interval, if no collisions have taken place in the meantime, the u, v, w components of velocity of each molecule will have increased respectively by amounts $X\,dt$, $Y\,dt$, $Z\,dt$,

---

* *Theory of Electrons*, Note 29, Teubner, Leipzig (1906) and David Nutt, London (1909).

† *Kinetische Theorie der Vorgänge in mässig verdünnten Gasen* (Inaug. Dissertation, Upsala, Almquist, 1917).

‡ *Vorlesungen über Gastheorie*, **1**, Chaps. II and III.

while the coordinates x, y, z will have increased respectively by amounts $u\,dt$, $v\,dt$, $w\,dt$.

It now follows, as in §22 above, that if no collisions occur, these molecules will have their velocity-components and space-coordinates lying within a small range $du\,dv\,dw\,dx\,dy\,dz$ surrounding the values $u+X\,dt$, etc., and $x+u\,dt$, etc. The number of molecules within this range at the end of the interval dt is, however,

$$vf(u+X\,dt,\ v+Y\,dt,\ w+Z\,dt,\ x+u\,dt,\ y+v\,dt,\ z+w\,dt,\ t+dt)$$
$$\times du\,dv\,dw\,dx\,dy\,dz, \quad \ldots\ldots(303)$$

and if no collisions occur, this expression must be exactly equal to expression (302). Expanding it as far as first powers of dt, and equating to expression (302), we obtain the relation

$$\frac{\partial}{\partial t}(vf) = -\left[X\frac{\partial}{\partial u}+Y\frac{\partial}{\partial v}+Z\frac{\partial}{\partial w}+u\frac{\partial}{\partial x}+v\frac{\partial}{\partial y}+w\frac{\partial}{\partial z}\right](vf),$$
$$\ldots\ldots(304)$$

expressing the rate at which vf changes on account of the motion of the molecules, and the forces acting on them.

When collisions occur, these produce an additional change in vf—let us say an increase at a rate

$$\left[\frac{\partial}{\partial t}(vf)\right]_{\text{coll.}}$$

per unit time. On combining the two causes of change in (vf), we arrive at the general equation

$$\frac{\partial}{\partial t}(vf) = -\left[X\frac{\partial}{\partial u}+Y\frac{\partial}{\partial v}+Z\frac{\partial}{\partial w}+u\frac{\partial}{\partial x}+v\frac{\partial}{\partial y}+w\frac{\partial}{\partial z}\right](vf)$$
$$+\left[\frac{\partial}{\partial t}(vf)\right]_{\text{coll.}}. \quad \ldots\ldots(305)$$

This equation must be satisfied by vf under all circumstances. When the gas is in a steady state the right-hand member must of course vanish.

No progress can be made with the development or solution of this equation until the term $\left[\dfrac{\partial}{\partial t}(vf)\right]_{\text{coll.}}$ has been evaluated, and this unfortunately can only be effected to a very limited extent. We saw in §§ 85, 86 that when the molecules are elastic spheres of diameter σ, this change is expressed by the equation

$$\left[\frac{\partial}{\partial t}(\nu f)\right]_{\text{coll.}} = \iiiint \nu^2(\bar{f}\bar{f}' - ff')\,V\sigma^2\cos\theta\,du'\,dv'\,dw'\,d\omega.$$

$$\ldots\ldots(306)$$

Let us now attempt to evaluate $\left[\dfrac{\partial}{\partial t}(\nu f)\right]_{\text{coll.}}$ when the molecules are regarded as point centres of force, attracting or repelling with a force which depends only on their distance apart.

We fix our attention on an encounter between two molecules, moving, before the encounter, with velocities u, v, w and u', v', w', and so with a relative velocity V, given by

$$V^2 = (u'-u)^2 + (v'-v)^2 + (w'-w)^2. \qquad\ldots\ldots(307)$$

In fig. 32 let O represent the centre of the first molecule moving in some direction QO with a velocity u, v, w, and let MNP represent the path described *relatively to* O by the second molecule before the encounter begins. When the second molecule comes to within such a distance of O that the action between the two molecules becomes appreciable, it will be deflected from its original rectilinear path MNP, and will describe a curved orbit such as MNS, this orbit being of course in the plane $MNPO$.

Let ROP be a plane through O perpendicular to MN, and let MN

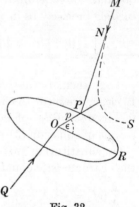

Fig. 32

meet this plane in a point P. Let the polar coordinates of P in the plane ROP be p, ϵ, the point O being taken as origin, and any line RO as initial line. Clearly p is the perpendicular from the first molecule O on to MN, the relative path of the second molecule before encounter.

Let us calculate the frequency of collisions in which the second molecule has a velocity u', v', w' whose components lie within a small specified range $du'\,dv'\,dw'$, while its path before the encounter is such that p, ϵ lie within a small range $dp, d\epsilon$. For such a collision the line MP must meet the plane ROP within a small element of area $p\,dp\,d\epsilon$. Thus the number of such collisions within an interval dt will be equal to the number of molecules which lie within a small volume $p\,dp\,d\epsilon\,V\,dt$, and have velocities within the specified range $du'\,dv'\,dw'$. This number is

$$\nu f(u',\,v',\,w')\,du'dv'dw'p\,dp\,d\epsilon\,V\,dt. \qquad \ldots\ldots(308)$$

The number of molecules per unit volume having velocities between u and $u+du$, v and $v+dv$, w and $w+dw$ is

$$\nu f(u,\,v,\,w)\,du\,dv\,dw,$$

so that the total number per unit volume of collisions of the kind we now have under consideration is

$$\nu^2 f(u,\,v,\,w)f(u',\,v',\,w')\,du\,dv\,dw\,du'dv'dw'p\,dp\,d\epsilon\,V\,dt. \qquad \ldots\ldots(309)$$

The type of collisions now under consideration is similar to that which we called type α in §85, and expression (309) is obviously a generalisation of our former expression (117). Proceeding precisely as in §86, we obtain, just as in formula (306),

$$\left[\frac{\partial}{\partial t}(\nu f)\right]_{\text{coll.}} = \iiiint \nu^2(\bar{f}\bar{f}' - ff')\,V\,du'dv'dw'p\,dp\,d\epsilon.$$
$$\ldots\ldots(310)$$

This, then, is the required generalisation of equation (306). It clearly reduces to this latter equation for elastic spheres, the factor $\sigma^2\cos\theta\,d\omega$ of equation (306) being exactly the factor $p\,dp\,d\epsilon$ of equation (310).

On substituting this value into equation (305), we obtain as the characteristic equation which must be satisfied by f,

$$\frac{\partial}{\partial t}(\nu f) = -\left[X\frac{\partial}{\partial u} + Y\frac{\partial}{\partial v} + Z\frac{\partial}{\partial w} + u\frac{\partial}{\partial x} + v\frac{\partial}{\partial y} + w\frac{\partial}{\partial z}\right](\nu f)$$
$$+ \iiiint \nu^2(\bar{f}\bar{f}' - ff')\,V\,du'dv'dw'p\,dp\,d\epsilon. \qquad \ldots\ldots(311)$$

For a mixture of gases, in which the different kinds of molecules are distinguished by the suffixes $1, 2, \ldots$, we obtain in a similar way a series of equations such as

$$\frac{\partial}{\partial t}(\nu_1 f_1) = -\left[X\frac{\partial}{\partial u} + Y\frac{\partial}{\partial v} + Z\frac{\partial}{\partial w} + u\frac{\partial}{\partial x} + v\frac{\partial}{\partial y} + w\frac{\partial}{\partial z}\right](\nu_1 f_1)$$
$$+ \Sigma \iiiint \nu_1\nu_2(\bar{f}_1\bar{f}'_2 - f_1 f'_2)\,V\,du'dv'dw'p\,dp\,d\epsilon. \qquad \ldots\ldots(312)$$

195. The condition for a steady state is that the right-hand members of equations (311) and (312) shall be equal to zero.

If we try putting

$$f = Ae^{-hm[(u-u_0)^2+(v-v_0)^2+(w-w_0)^2]}, \qquad \ldots\ldots(313)$$

the integrals on the right of equations (311) and (312) vanish, so that this value of f provides a solution when $X = Y = Z = 0$, and νf is independent of x, y, z. It is of course the solution already found in § 87.

On substituting

$$f = Ae^{-hm(c^2+2\chi)}, \qquad \ldots\ldots(314)$$

we again obtain a solution, provided

$$X = -\frac{\partial \chi}{\partial x}, \quad Y = -\frac{\partial \chi}{\partial y}, \quad Z = -\frac{\partial \chi}{\partial z}.$$

Thus (314) is a solution when χ is the potential of the forces acting on the molecule, the result obtained in § 23.

If, however, u_0, v_0, w_0, h and ν vary from point to point, formulae (313) and (314) do not provide a solution, for on substituting them into the right-hand member of equation (311) we find that the second term vanishes, while the first does not.

The case in which u_0, v_0, w_0, h and ν vary only slightly from point to point is especially important for the study of viscosity, conduction of heat and diffusion. In this case we may assume, as an appropriate solution

$$f = f_0[1 + \Phi(x, y, z, u, v, w)], \qquad \ldots\ldots(315)$$

where Φ is a small quantity of the first order, and

$$f_0 = Ae^{-hm[(u-u_0)^2+(v-v_0)^2+(w-w_0)^2]}. \qquad \ldots\ldots(316)$$

Since $f = f_0$ is a solution when u_0, v_0, w_0, h and ν do not vary from point to point, equation (315) must necessarily provide a solution when these quantities vary to the first order of small quantities.

The right-hand member of equation (312) contains the term $\nu_1\nu_2 f_1 f_2'$ of which the value, by equation (315), is

$$\nu_1\nu_2 f_1 f_2' = \nu_1\nu_2 f_{01} f_{02}'(1 + \Phi_1 + \Phi_2'). \qquad \ldots\ldots(317)$$

Here f_{01} denotes the value of f_0 for a molecule of the first kind, and so on, and the product $\Phi_1\Phi_2'$ is omitted as being of the second order of small quantities. Similarly

$$\nu_1\nu_2\bar{f}_1\bar{f}_2' = \nu_1\nu_2\bar{f}_{01}\bar{f}_{02}'(1+\bar{\Phi}_1+\bar{\Phi}_2'). \qquad \ldots\ldots(318)$$

The conservation of energy and momenta at an encounter give the relation

$$\bar{f}_{01}\bar{f}_{02}' = f_{01}f_{02}', \qquad \ldots\ldots(319)$$

so that $\quad \nu_1\nu_2(\bar{f}_1\bar{f}_2' - f_1f_2') = \nu_1\nu_2 f_{01}f_{02}'(\bar{\Phi}_1+\bar{\Phi}_2'-\Phi_1-\Phi_2')$.

On substituting the tentative solution (315) into equation (312), f may be replaced by f_0 everywhere except in the integrals; for if we retained terms in Φ in the remaining parts of the equation, we should be including terms of the second order of small quantities. Equation (312) accordingly reduces to

$$\left(\frac{\partial}{\partial t}+X\frac{\partial}{\partial u}+Y\frac{\partial}{\partial v}+Z\frac{\partial}{\partial w}+u\frac{\partial}{\partial x}+v\frac{\partial}{\partial y}+w\frac{\partial}{\partial z}\right)(\nu_1 f_{01}) = \nu_1 f_{01}I,$$
$$\ldots\ldots(320)$$

where

$$I = \Sigma\iiiint\int\int\nu_2 f_{02}'(\Phi_1+\bar{\Phi}_2'-\Phi_1-\Phi_2')\,V\,du'dv'dw'p\,dp\,d\epsilon,$$
$$\ldots\ldots(321)$$

an equation in which every term is of the first order of small quantities.

On dividing out by $\nu_1 f_{01}$ and replacing f_{01} by its value from equation (316), this equation becomes

$$\left(\frac{\partial}{\partial t}+X\frac{\partial}{\partial u}+Y\frac{\partial}{\partial v}+Z\frac{\partial}{\partial w}+u\frac{\partial}{\partial x}+v\frac{\partial}{\partial y}+w\frac{\partial}{\partial z}\right)$$
$$\times\{\log\nu_1 A - hm[(u-u_0)^2+(v-v_0)^2+(w-w_0)^2]\} = I.$$
$$\ldots\ldots(322)$$

The solution (315) is indeterminate in the sense that changes in u_0, v_0, w_0, νA and h are not independent of changes in Φ. For instance, the total momentum in the direction Ox is readily found to be

$$m\nu u_0 + m\iiint\nu f_0\,\Phi u\,du\,dv\,dw \qquad \ldots\ldots(323)$$

and the same change can be made in this either by changing u or by changing Φ.

Let us agree that u_0, v_0, w_0, νA and h are to have the same physical interpretation in solution (315) as they have in the steady state solution for which $\Phi = 0$, e.g. the velocity of mass-motion of the gas is to have components u_0, v_0, w_0, and so on. This makes the value of Φ quite definite for any assigned physical conditions. But for this condition to be satisfied, the second term in expression (323) must vanish, with similar equations in v and w, and Φ must also satisfy the equations

$$\iiint \nu f_0 \Phi \, du \, dv \, dw = 0, \qquad \ldots \ldots (324)$$

$$\iiint \nu f_0 \Phi (u^2 + v^2 + w^2) \, du \, dv \, dw = 0. \qquad \ldots \ldots (325)$$

With Φ restricted in this way, the equation of continuity is again given by equation (285), so that

$$\frac{d}{dt} \log \nu = -\left(\frac{\partial u_0}{\partial x} + \frac{\partial v_0}{\partial y} + \frac{\partial w_0}{\partial z} \right), \qquad \ldots \ldots (326)$$

in which small quantities of the second order, such as $u_0 \dfrac{\partial \nu}{\partial x}$ are neglected. We now multiply both sides of this equation by $\frac{2}{3} h m c^2$, add to the corresponding sides of equation (322), and find that the latter equation reduces to

$$(1 + \tfrac{2}{3} h m c^2) \frac{\partial}{\partial t} \log \nu + \frac{\partial}{\partial t} \{ \tfrac{3}{2} \log h - h m [(u - u_0)^2$$

$$+ (v - v_0)^2 + (w - w_0{}^2] \} - 2 h m [(u - u_0) X + (v - v_0) Y + (w - w_0) Z]$$

$$+ \left(u \frac{\partial}{\partial x} + v \frac{\partial}{\partial y} + w \frac{\partial}{\partial z} \right) \log \nu + 2 h m \left[(u^2 - \tfrac{1}{3} c^2) \frac{\partial u_0}{\partial x} + (v^2 - \tfrac{1}{3} c^2) \frac{\partial v_0}{\partial y} \right.$$

$$\left. + (w^2 - \tfrac{1}{3} c^2) \frac{\partial w_0}{\partial z} + uv \left(\frac{\partial v_0}{\partial x} + \frac{\partial u_0}{\partial y} \right) + \cdots \right]$$

$$- \left(m c^2 - \frac{3}{2h} \right) \left(u \frac{\partial h}{\partial x} + v \frac{\partial h}{\partial y} + w \frac{\partial h}{\partial z} \right) = I, \qquad \ldots \ldots (327)$$

where I is given by equation (321).

It will be remembered that this equation is only accurate when Φ satisfies five relations, expressed by equations (323), (324) and (325). The solutions in Φ will however be additive, since the equations are linear; five solutions which contribute nothing to either side are

$$\Phi = 1,\, mu,\, mv,\, mw,\, mc^2,$$

so that to any solution for Φ which satisfies equation (327) may be added terms of the form

$$\Phi = B + Cmu + Dmv + Emw + Fmc^2,$$

and the constants $B,\, C,\, D,\, E,\, F$ may be adjusted so as to satisfy the five necessary conditions.

Law of Force μr^{-s}

196. The next step must be to calculate I, as given by equation (321), and we can only do this by assuming definite laws for the interaction between molecules at collisions. We therefore suppose that the molecules are centres of force repelling according to the law μr^{-s}.*

If two molecules of masses m_1, m_2 at a distance r apart exert a repulsive force of this amount, then it is easily shewn that the motion of either relative to the other is that of a particle of unit mass moving about a fixed centre of force, the potential energy when the two are at a distance r apart being

$$\frac{(m_1 + m_2)\,\mu}{m_1 m_2 (s-1)\, r^{s-1}}.$$

As usual, the differential equation of the orbit is

$$\frac{1}{2}\frac{h^2}{r^4}\left\{\left(\frac{\partial r}{\partial \theta}\right)^2 + r^2\right\} = C - \frac{(m_1 + m_2)\,\mu}{m_1 m_2 (s-1)\, r^{s-1}},$$

where r, θ are now polar coordinates in the plane of the relative orbit, h is the angular momentum pV, and $C = \frac{1}{2}V^2$, where V, the relative velocity before the encounter begins, is identical with the usual "velocity at infinity".

This equation has the integral

$$\theta = \int_{\infty}^{r} \frac{dr}{\sqrt{\dfrac{2C}{h^2}r^4 - r^2 - \dfrac{2(m_1 + m_2)\,\mu}{m_1 m_2 (s-1)\, h^2}r^{5-s}}},$$

in which the addition of a constant of integration is avoided by

* The method of §196 is that of Maxwell, *Collected Works*, **2**, p. 36.

taking the direction of the asymptote to the orbit to be the initial
line $\theta = 0$.

Replacing h by pV, and further writing η for p/r, so that η is a
pure number, this becomes

$$\theta = \int_0^{\eta} \frac{d\eta}{\sqrt{1 - \eta^2 - \dfrac{2}{s-1}\left(\dfrac{\eta}{\alpha}\right)^{s-1}}}, \qquad \ldots\ldots(328)$$

where $\qquad \alpha = p\left(\dfrac{m_1 m_2 V^2}{(m_1 + m_2)\,\mu}\right)^{\frac{1}{s-1}}. \qquad \ldots\ldots(329)$

The apses of the orbit are given by $\partial r/\partial \theta = 0$, and therefore by

$$1 - \eta^2 - \frac{2}{s-1}\left(\frac{\eta}{\alpha}\right)^{s-1} = 0.$$

A simple graphical treatment shews that, when s is greater than
unity, this equation can only have one real root. Call this root
η_0, then the angle, say θ_0, between the asymptote and the apsidal
distance will be given by equation (328) on taking the upper limit
to be η_0. The angle between the asymptotes, say θ', is equal to
twice this, and so is given by

$$\theta' = 2\theta_0 = 2\int_0^{\eta_0} \frac{d\eta}{\sqrt{1 - \eta^2 - \dfrac{2}{s-1}\left(\dfrac{\eta}{\alpha}\right)^{s-1}}}.$$

After the encounter, the velocities parallel and perpendicular to the initial line
are of course $-V\cos\theta'$ and $-V\sin\theta'$.

For any value of s, there will naturally
be a doubly infinite series of possible orbits
corresponding to different values of p and
V. Except for a difference of linear scale,
however, these may be reduced to a singly
infinite system corresponding to the varia-
tion of α or $pV^{\frac{2}{s-1}}$. Fig. 33* shews some
members of this singly infinite system for
the law of force μ/r^5.

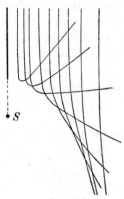

Fig. 33

* This figure is given by Maxwell, *Collected Works*, **2**, p. 42. I am indebted to the Cambridge University Press for the use of the original block.

In fig. 32 of § 194, the angle ϵ was measured from an arbitrarily chosen line OR. Let us now suppose this to be the intersection of the plane POR with a plane through O containing the direction of NP and the axis of x, as in fig. 34.

In fig. 35, let OR, OX be the directions of this line OR and of the axis of x. Let OG be the direction of V, the relative velocity before collision, so that OR, OX, OG all lie in one plane. Let these lines be supposed each of unit length, so that the points GXR lie on a sphere of unit radius about O as centre.

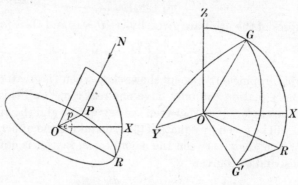

Fig. 34   Fig. 35

Let OY, OZ be unit lines giving the directions of the axes of y and z, and let OG' give the direction of the relative velocity after the encounter. Then GOG' is the plane of the orbit, which is the plane NPO in fig. 34. Thus the angle RGG' is the ϵ of § 194, while the angle GOG' is θ'.

Now suppose that, as usual, the velocities before collision are u, v, w and u', v', w', so that the relative velocity V is given by

$$V^2 = (u'-u)^2 + (v'-v)^2 + (w'-w)^2,$$

and the velocities after collision are $\overline{u}, \overline{v}, \overline{w}$ and $\overline{u'}, \overline{v'}, \overline{w'}$.

From the spherical triangle $G'GX$,

$$\cos G'X = \cos GX \cos GG' + \sin GX \sin GG' \cos \epsilon,$$

in which we have $\cos G'X = -\dfrac{\overline{u'}-\overline{u}}{V}$, $\cos GX = \dfrac{u'-u}{V}$, so that

$$\overline{u}-\overline{u'} = (u'-u)\cos \theta' + \sqrt{V^2-(u'-u)^2}\sin \theta' \cos \epsilon. \quad ......(330)$$

Denoting the angle XGY by ω_2 and XGZ by ω_3, we have, in a similar way,

$$\overline{v} - \overline{v'} = (v' - v)\cos\theta' + \sqrt{V^2 - (v' - v)^2}\sin\theta'\cos(\epsilon - \omega_2),$$

$$\overline{w} - \overline{w'} = (w' - w)\cos\theta' + \sqrt{V^2 - (w' - w)^2}\sin\theta'\cos(\epsilon - \omega_3).$$

To determine ω_2, we notice that in the triangle GXY, $XY = \frac{1}{2}\pi$ and $XGY = \omega_2$; thus

$$(u' - u)(v' - v) + \sqrt{[V^2 - (u' - u)^2][V^2 - (v' - v)^2]}\cos\omega_2 = 0,$$

$$(u' - u)(w' - w) + \sqrt{[V^2 - (u' - u)^2][V^2 - (w' - w)^2]}\cos\omega_3 = 0.$$

These equations together with three equations of momentum, such as

$$m_1\overline{u} + m_2\overline{u'} = m_1 u + m_2 u', \qquad \ldots\ldots(331)$$

determine the velocities after collision.

Eliminating $\overline{u'}$ from equations (330) and (331), we obtain

$$\overline{u} = u + \frac{m_2}{m_1 + m_2}[2(u' - u)\cos^2\tfrac{1}{2}\theta' + \sqrt{V^2 - (u' - u)^2}\sin\theta'\cos\epsilon],$$

$$\ldots\ldots(332)$$

and there are of course similar equations giving $\overline{v}$ and $\overline{w}$.

Solutions for Φ

197. We are now in a position to proceed with the evaluation of I and the discussion of equation (327). From equation (329) we have

$$p\,dp\,d\epsilon = [(m_1 + m_2)\mu/m_1 m_2]^{\frac{2}{s-1}} V^{-\frac{4}{s-1}}\alpha\,d\alpha\,d\epsilon,$$

so that

$$I = \Sigma\nu_2[(m_1 + m_2)\mu/m_1 m_2]^{\frac{2}{s-1}} J,$$

where J stands for the quintuple integral

$$J = \iiiint\int (\overline{\Phi}_1 + \overline{\Phi}'_2 - \Phi_1 - \Phi'_2)\,\alpha\,d\alpha\,d\epsilon\, V^{\frac{s-5}{s-1}} f'_{02}\,du'\,dv'\,dw'.$$

$$\ldots\ldots(333)$$

In this integral, it will be remembered that Φ_1 is a function, as yet undetermined, of u, v and w; Φ'_2 is the similar function of

u', v', w' for a molecule of the second kind, while $\bar{\Phi}_1$, $\bar{\Phi}_2'$ have corresponding meanings in terms of the velocities after collision. Our problem is not to evaluate expression (333) for given values of Φ, but to find values of Φ such that after integration expression (333) shall be equal to a certain algebraic function containing terms of degrees 0, 1, 2 and 3 in u, v and w, namely that on the left of equation (327).

We see at once the simplification introduced by supposing, with Maxwell, that $s = 5$. In this special case, the power of V disappears entirely from the integral J.

In the more general case, as discussed by Lorentz[*] and Enskog,[†] we consider a tentative solution for equation (327) of the form

$$\Phi = u\phi_1(c).$$

With this value for Φ, let the value of J as defined by equation (333) be denoted by

$$J = J(u),$$

where $J(u)$, in addition to depending on u, depends on the various constants of the gas, m_1, m_2, h, etc. Similarly, let $\Phi = v\phi_1(c)$ lead to a value $J = J(v)$ and let $\phi = w\phi_1(c)$ lead to a value $J = J(w)$.

Then a solution

$$\Phi = (lu + mv + nw)\,\phi_1(c) \qquad \ldots\ldots(334)$$

will clearly lead to

$$J = lJ(u) + mJ(v) + nJ(w). \qquad \ldots\ldots(335)$$

Now let l, m, n be regarded as direction-cosines, so that $lu + mv + nw$ will be the component velocity along a certain direction (l, m, n); it follows that the solution (334) will lead to a value of J given by

$$J = J(lu + mv + nw), \qquad \ldots\ldots(336)$$

where $J(lu + mv + nw)$ is the same function of $lu + mv + nw$ as $J(u)$ is of u. Comparing (335) and (336), we have

$$J(lu + mv + nw) - lJ(u) - mJ(v) - nJ(w) = 0,$$

* *Vorträge über die Kinetische Theorie der Materie und der Elektrizität* (Leipzig, 1914), p. 185.

† *Kinetische Theorie der Vorgänge in mässig verdünnten Gasen* (Inaug. Dissertation, Upsala, 1917).

whence it can be shewn that $J(u)$ must be of the form

$$J(u) = u\chi_1(c), \text{ etc.}$$

Thus we have shewn that a solution

$$\Phi = u\phi_1(c)$$

leads to a value of J of the form

$$J = u\chi_1(c).$$

By a somewhat similar method, Enskog shews* that a solution

$$\Phi = (u^2 - \tfrac{1}{3}c^2)\,\phi_2(c)$$

leads to a value of J of the form

$$J = (u^2 - \tfrac{1}{3}c^2)\,\chi_2(c),$$

while a solution $\qquad \Phi = uv\phi_2(c)$

leads to $\qquad\qquad J = uv\chi_2(c).$

198. Thus the solution of the general steady-state equation will be

$$\Phi = \psi_1 + \psi_2 + \psi_3 + \psi_4 + B + Cmu + Dmv + Emw + Fmc^2,$$

$$\dots\dots(337)$$

where ψ_1, ψ_4 have the forms

$$\psi_1 = u\left\{2hmX - \frac{1}{\nu}\frac{\partial\nu}{\partial x} + \left(mc^2 - \frac{3}{2h}\right)\frac{\partial h}{\partial x}\right\}\phi_1(c),$$

$$\psi_4 = \left\{(u^2 - \tfrac{1}{3}c^2)\frac{\partial u_0}{\partial x} + \dots + uv\left(\frac{\partial v_0}{\partial x} + \frac{\partial u_0}{\partial y}\right) + \dots\right\}\phi_2(c),$$

and ψ_2, ψ_3 are obtained from ψ_1 by changing x into y, z respectively, u into v, w and X into Y, Z.

Here $\phi_1(c)$, $\phi_2(c)$ are functions of c and the constants of the gas only. As we have already seen, they cannot be evaluated in finite terms. Their expansion for a series of powers of c^2 has been considered by Enskog,† who has also calculated numerical results.‡

We are now in a position to apply the analysis of the present chapter to the exact solution of physical problems.

* Enskog, *l.c.* pp. 39, 40. † *L.c.* Chaps. II and III.
‡ *L.c.* Chap. IV.

Viscosity

199. In discussing viscosity, the gas may be supposed to be at the same temperature throughout, so that

$$\frac{\partial h}{\partial x} = \frac{\partial h}{\partial y} = \frac{\partial h}{\partial z} = 0,$$

and at uniform pressure throughout, so that

$$\frac{\partial v}{\partial x} = \frac{\partial v}{\partial y} = \frac{\partial v}{\partial z} = 0.$$

We may also suppose that no external forces are acting on the gas so that

$$X = Y = Z = 0.$$

As a consequence of this, ψ_1, ψ_2, ψ_3 are all equal to zero, and the solution (337) reduces to

$$\Phi = \psi_4.$$

The law of distribution now becomes (cf. equation (315))

$$f = A e^{-hm(u^2+v^2+w^2)}(1 + \psi_4).$$

The pressures can now be calculated from equations (292). Neglecting the intermolecular pressures ϖ_{xx}, etc., these are found to be of the form

$$P_{xx} = p - 2\eta\, \frac{\partial u_0}{\partial x} + \frac{2}{3}\,\eta\left(\frac{\partial u_0}{\partial x} + \frac{\partial v_0}{\partial y} + \frac{\partial w_0}{\partial z}\right), \quad \ldots\ldots(338)$$

$$P_{xy} = -\eta\left(\frac{\partial v_0}{\partial x} + \frac{\partial u_0}{\partial y}\right), \quad\quad\quad\quad \ldots\ldots(339)$$

where η can only be calculated after $\phi_1(c)$ has been calculated. These are, however, precisely the equations of motion of a viscous fluid having a coefficient of viscosity η.[*] Thus the pressures given by equations (338) and (339) will be exactly accounted for by regarding the gas as a viscous fluid having a coefficient of viscosity η.

The value of η has been calculated by Enskog,[†] who found the values already quoted in Chap. VI.

---

[*] Lamb, *Hydrodynamics*, p. 512.   [†] *L.c.* pp. 88 ff.

Conduction of Heat

200. Consider next a gas which is in a steady state, and devoid of mass-motion, but which is not at a uniform temperature. For simplicity suppose that the gas is arranged in parallel strata of equal temperature, so that the temperature is a function of z only, and that no external forces act, so that $X = Y = Z = 0$.

In the general solution (337) of the steady state equation, we now have $\psi_2 = \psi_3 = \psi_4 = 0$, so that

$$\varPhi = \psi_1 = u\left\{-\frac{1}{\nu}\frac{\partial \nu}{\partial x} + \left(mc^2 - \frac{3}{2h}\right)\frac{\partial h}{\partial x}\right\}\phi_1(c).$$

The relation between $\dfrac{\partial \nu}{\partial x}$ and $\dfrac{\partial h}{\partial x}$ must be found from the condition that there shall be no mass-motion, i.e. the process of conduction must not be complicated by the addition of convection. To satisfy this condition, we must have

$$\iiint \nu f_0\, u\psi_1\, du\, dv\, dw = 0.$$

On substituting for ψ_1 in this equation, and carrying out the integrations, we obtain a linear relation between $\dfrac{\partial \nu}{\partial x}$ and $\dfrac{\partial h}{\partial x}$. Using this relation, $\varPhi$ is given as a multiple of $\dfrac{\partial h}{\partial x}$, or of

$$-\frac{1}{2hT^2}\frac{\partial T}{\partial x},$$

and from this it is easy to calculate the coefficient of conduction of heat, by the methods already used in § 199. The result is that given in § 155.

Diffusion

201. To discuss the phenomenon of diffusion, we may in the first place suppose the gas to be at a uniform temperature throughout, and further take u_0, v_0, w_0 all constant and X, Y, Z all zero. A sufficient solution of equation (337) is now seen to be

$$\varPhi = \psi_1 = \frac{u}{\nu}\frac{\partial \nu}{\partial x}\phi_1(c),$$

corresponding to diffusion in the direction of the axis of x. The flow of molecules parallel to the axis of x, measured per unit area per unit time, is

$$\iiint \nu \left(\frac{hm}{\pi}\right)^{\frac{3}{2}} e^{-hmc^2} \left(u - \frac{u^2}{\nu}\frac{\partial \nu}{\partial x}\phi_1(c)\right) du\,dv\,dw.$$

The coefficient of diffusion is the coefficient of $-\partial \nu/\partial x$ in this expression, and is found to have the values already quoted in §172.

Maxwell's treatment of the diffusion problem was based upon his general equation of transfer (299), which assumes the form

$$\frac{\partial}{\partial x}(\nu \overline{uQ}) - \overline{Q}\frac{\partial}{\partial x}(\nu \overline{u}) = \varDelta Q, \qquad \ldots\ldots(340)$$

for steady motion parallel to the axis of x. On putting $Q = u$ and assuming the inverse fifth-power law, this equation becomes

$$\frac{1}{2hm_1}\frac{\partial \nu_1}{\partial x} = \nu_1 \nu_2 m_2 \sqrt{\frac{\mu}{m_1 m_2(m_1 + m_2)}} A_1(u_{02} - u_{01}), \quad (341)$$

where A_1 is the constant introduced and defined in §172, and a comparison with the general equation of diffusion enabled Maxwell to determine the coefficient of diffusion.

The success of the method depended on the assumption of the law of the inverse fifth-power. Under this law $\varDelta u$ was found to be proportional to $u_{02} - u_{01}$ and the equation of diffusion followed at once. Under any other law $\varDelta u$ is not proportional to $u_{02} - u_{01}$; the value of $\varDelta Q$ depends on the law of distribution of velocities whatever value is given to Q, and the equation of diffusion no longer follows on putting $Q = u$.

To obtain the equation of diffusion from equation (340) in the general case, it is necessary to assume Q to be equal to u multiplied by a series of powers of c. The resulting equation* is found to be of the same general type as (341), except for the important

---

* For details see Chapman, *Phil. Trans.* **217** A (1917), pp. 124, 181. The analysis given by Chapman in this paper is invalidated by an error of algebra, as was pointed out by Enskog (*Arkiv f. Mat. Astron. und Fysik*, **16** (1921)), but the numerical consequences are not serious. For a corrected discussion see Chapman and Hainsworth, *Phil. Mag.* **48** (1924), p. 593.

difference that the left-hand member includes terms in $\partial T/\partial x$ and $\partial p/\partial x$ in addition to the term in $\partial v/\partial x$.

If T and p do not vary with x, these additional terms disappear. But their presence in the general equation indicates that a process of diffusion is necessarily going on in any gas in which T and p vary from point to point, exception being made of the special case in which the molecules repel according to the exact inverse fifth-power of the distance. These phenomena were first predicted on purely theoretical grounds by Enskog* in 1917; shortly afterwards they were predicted independently by Chapman in the paper already quoted.

Thermal- and Pressure-diffusion

202. The phenomenon of "pressure-diffusion" originates in the terms in $\partial p/\partial x$, but does not appear to possess any great importance physically. That of "thermal-diffusion", which originates in the terms in $\partial T/\partial x$, is of importance because numerically its magnitude is comparable with that of ordinary diffusion. Let us imagine that we have a tube or cylinder, originally filled with a uniform mixture of two gases, and let the two ends be kept permanently at different temperatures. As the result of thermal diffusion, currents will be set up in the tube, the molecules of the heavier gas tending to diffuse in the direction of decreasing temperature and vice versa. There is a limit to the inequality of composition of the mixture which can be established by this means, for the inequality brings into play ordinary diffusion which acts in the opposite direction and tends to restore uniformity of composition. Thus a steady state will ultimately be reached in which the proportion of the mixture will vary gradually as we pass along the tube.

This predicted variation in the proportion of the mixture was first confirmed experimentally by F. W. Dootson.† In a typical experiment, a tube with a bulb at each end was filled with a mixture of hydrogen and carbon-dioxide in approximately equal proportions. One bulb was then kept for four hours at a steady

---

* *Kinetische Theorie der Vorgänge in mässig verdünnten Gasen* (Inaug. Dissertation, Upsala, 1917).

† *Phil. Mag.* **33** (1917), p. 248.

temperature of 230° C., the other being kept water-cooled at 10° C. At the end of the four hours samples were drawn off from the two bulbs and analysed, with the following results:

Hot bulb (230° C.): 44·9 per cent H_2; 55·1 per cent CO_2,

Cold bulb (10° C.): 41·3 per cent H_2; 58·7 per cent CO_2.

The effect detected here is of the sign and order of magnitude predicted by theory. In actual amount it is rather less than half the amount predicted on the supposition that the molecules behave like elastic spheres. The theoretical effect for elastic spheres is, however, greater than that for any other type of molecule, and it will be remembered that it vanishes altogether for molecules repelling according to the Maxwellian law μr^{-5}. Thus the effect detected by experiment was about what was to be expected, and as later and more accurate experiments* have given very similar results, the theory may be regarded as being verified.†

The steady state phenomenon we have been considering is of interest in that it depends greatly upon the law of force between molecules. It seems possible that it may in time lead to powerful methods for the investigation of molecular fields of force.

Chapman has shown how it can be used for the separation of gases which have the same molecular weight (e.g. C_2H_4 and N_2), and it has recently proved useful for the separation of isotopes.

* Ibbs, *Proc. Roy. Soc.* **99** A (1921), p. 385, and **107** A (1925), p. 470; Elliott and Masson, *Proc. Roy. Soc.* **108** A (1925), p. 378.

† A fuller discussion of both the theory and the experimental verification of thermal diffusion will be found in Chapman and Cowling, *The Mathematical Theory of Non-uniform Gases* (C. U. Press, 1939), pp. 252–258.

Chapter X

GENERAL STATISTICAL MECHANICS
AND THERMODYNAMICS

203. So far our molecules have been treated either as elastic
spheres, exerting no forces on one another except when in actual
collision, or else as point centres of force, attracting or repelling
according to comparatively simple laws. The time has now come
to discard all such restrictions, and treat the question in a more
general way, regarding the molecules as general mechanical
structures, which may be as complicated as we please, consisting
of any number of parts, capable of any kind of internal motion
and exerting upon one another forces of any type.

Degrees of Freedom

204. The total number of independent quantities which are
needed to specify the configuration of any mechanical system is
called the number of degrees of freedom of the system. This
number does not depend on the motions, but on the capacities
for motion, of the various parts of the system; it is therefore
related to the geometrical or kinematical, and not to the
mechanical, properties of the system.

For example, if a point is free to move in space, its position can
be specified by three quantities, as for instance x, y, z, the rectangular coordinates of the point, so that a point which is free to
move in space has three degrees of freedom. A rigid body which
is free to move in space has six degrees of freedom, for the position
of the body can only be fully fixed when six quantities are known,
as for instance x, y, z the coordinates of the centre of gravity of the
body, and three angles to determine the orientation of the body.
Similarly, a pair of compasses has seven degrees of freedom, two
rigid arms connected by a "universal joint" has nine, and so on.

Numbers of degrees of freedom are additive in the sense that
when a complex system is made up of a number of simpler
systems, the number of degrees of freedom of the complex system

is equal to the sum of the numbers of degrees of freedom of the constituent systems. We see this at once on noticing that a knowledge of the configuration of the complex system is exactly equivalent to a knowledge of the configurations of all the constituent systems.

If atoms are regarded as points, each atom has three degrees of freedom. A diatomic molecule must therefore have six degrees of freedom. These can be counted up in a variety of different ways, but the total must always come to six. For instance, we might take the six degrees of freedom to consist of the three degrees of freedom of the centre of gravity of the molecule to move in space, the two degrees of freedom of the line joining the two atoms to change its direction in space, and the one degree of freedom arising from the possibility of the two atoms changing their distance apart. Alternatively, we may suppose each atom to have its own three degrees of freedom as a point, so that the diatomic molecule again has six degrees of freedom.

In general, if atoms are regarded as points, a molecule composed of n atoms will have $3n$ degrees of freedom, but if atoms are regarded as rigid bodies capable of rotational as well as translational motion, such a molecule will have $6n$ degrees of freedom. If electrons are regarded as points, a cluster of n-electrons will have $3n$ degrees of freedom. If molecules are treated as points, a gas consisting of N molecules will have $3N$ degrees of freedom, while if each molecule has n degrees of freedom, the gas will have nN degrees of freedom.

The Motion of a General Dynamical System

205. A dynamical system which has n degrees of freedom needs n coordinates to specify its configuration. Let us call these

$$q_1, q_2, \ldots q_n, \qquad \ldots\ldots(342)$$

and let us denote their rates of increase by

$$\dot{q}_1, \dot{q}_2, \ldots \dot{q}_n. \qquad \ldots\ldots(343)$$

The configuration and motion of the system at any instant are completely known when the values of the above $2n$ quantities are known, and classical mechanics tells us that, if these quantities are known at any instant, it is possible to trace out the motion of the system throughout all future time, or until it experiences some disturbance from outside. The energy of the system depends only on these $2n$ quantities; let us denote it by E. It is convenient to introduce quantities known as momenta, which we denote by

$$p_1, \, p_2, \, \ldots \, p_n$$

and define by the equations

$$p_s = \frac{\partial E}{\partial \dot{q}_s}, \text{ etc.}$$

Ordinarily E will contain only squares and products of the velocities $\dot{q}_1, \dot{q}_2, \ldots \dot{q}_n$, so that the momenta are linear functions of these velocities.

For instance, if the dynamical system is simply a moving particle of mass m, its coordinates may be supposed to be x, y, z, its velocities are $\dot{x}, \dot{y}, \dot{z}$, its energy is

$$E = \tfrac{1}{2}m(\dot{x}^2 + \dot{y}^2 + \dot{z}^2),$$

so that the momenta are

$$\frac{\partial E}{\partial \dot{x}} = m\dot{x}, \text{ etc.}$$

In this case the momenta reduce to the ordinary momenta of a moving mass, mu, mv, mw.

A simplification is introduced on replacing the $2n$ quantities (342) and (343) by the $2n$ quantities

$$q_1, \, q_2, \, \ldots \, q_n, \, p_1, \, p_2, \, \ldots \, p_n. \qquad \ldots\ldots(344)$$

When the values of these quantities are known at any instant, the configuration and velocities of the system are known at that instant, so that the classical mechanics tells us that the motion of the system can be traced throughout all time.

The equations which determine this motion are the well-known Hamiltonian equations

$$\frac{dp_s}{dt} = -\frac{\partial E}{\partial q_s}; \quad \frac{dq_s}{dt} = \frac{\partial E}{\partial p_s}, \qquad \ldots\ldots(345)$$

in which it is essential that E is expressed as a function of the q's and p's.

The Conception of a Generalised Space

206. Twice already we have found it convenient to represent the three components of velocity u, v, w of a molecule by a point in space of coordinates u, v, w. The same artifice is found useful on many occasions. And, although the space of nature possesses only three dimensions, we need not limit our use of the artifice to spaces of three dimensions, for the spaces we use exist only in our imagination. If the configuration and motion of a dynamical system are specified by the $2n$ quantities (344), we can imagine these quantities represented as the coordinates of a space of $2n$ dimensions. Such a space is called a "phase-space".

Each point in this space represents one set of values of the $2n$ coordinates (344), and so represents one state of the system under discussion. As the natural motion of the system proceeds, the point moves in the phase-space, and its motion records the history of the system. The rates at which the point moves in the different directions of the phase-space—i.e. the components of velocity of the point—are given by equations (345). The ratios of these components give the direction of motion at the point, and we see that this depends only on the coordinates of the point in the phase-space. Thus the moving points move along curves in the phase-space, the positions of which do not change with the time. These curves are generally known as "trajectories".

We can imagine all such trajectories mapped out in the phase-space, and the $2n$-dimensional chart so obtained will enable us to follow out the motion of our dynamical system, starting from any initial state that we please. It is as though the currents in the Atlantic Ocean were always the same at each point of the Ocean. A chart could then be drawn, like the usual Admiralty chart, which would shew the motion of the surface-water at each point of the Ocean. Such a chart would enable us to trace the motion of each particle of surface-water throughout its whole life in the Ocean.

The problem before us is not, however, to study the motion of our system when it starts from any particular set of initial conditions; we wish rather to find statistical properties which shall be true for its motions, no matter from what particular state it starts. We therefore imagine the whole of the $2n$-dimensional space filled with moving points each describing its own curve, as determined by equations (345). We shall suppose these points to be so thickly scattered in the space that they may be regarded as forming a continuous fluid; our object is now to make a statistical study of the particles of this fluid. Let us first consider the motion of the points which originally occupy a small rectangular element of volume in the phase-space extending from p_1 to $p_1 + dp_1$, p_2 to $p_2 + dp_2$, etc., and so of content dv equal to $dp_1, dp_2, \ldots dq_n$.

Let us follow these points in their motion through a small interval of time dt. Those which originally occupy the face p_1 of the small element of volume are moving perpendicular to the face at a rate $\dot{p}_1$, while those in the opposite face $p_1 + dp_1$ are moving at a rate $\dot{p}_1 + \dfrac{\partial \dot{p}_1}{\partial p_1} dp_1$. Thus these two layers of particles are moving apart at a rate $\dfrac{\partial \dot{p}_1}{\partial p_1} dp_1$, and by the end of the time dt their distance apart will have increased from dp_1 to

$$\left(1 + \frac{\partial \dot{p}_1}{\partial p_1} dt\right) dp_1.$$

There is a similar increase in the distance apart of the slabs of particles which formed each other pair of faces, so that after an interval dt the points under consideration will occupy a volume

$$\left(1 + \frac{\partial \dot{p}_1}{\partial p_1} dt\right)\left(1 + \frac{\partial \dot{p}_2}{\partial p_2} dt\right) \ldots dp_1 dp_2 \ldots.$$

Multiplying out and neglecting squares of dt, this becomes

$$\left[1 + \left(\frac{\partial \dot{p}_1}{\partial p_1} + \frac{\partial \dot{p}_2}{\partial p_2} + \ldots + \frac{\partial \dot{q}_1}{\partial q_1} + \frac{\partial \dot{q}_2}{\partial q_2} + \ldots\right) dt\right] dp_1 dp_2 \ldots \quad \ldots \ldots (346)$$

The values of $\dot{p}_1, \dot{q}_1$, etc. are of course given by equations (345). From these we at once find that

$$\frac{\partial \dot{p}_1}{\partial p_1} + \frac{\partial \dot{q}_1}{\partial q_1} = 0,$$

and similarly for every other pair of coordinates, so that the coefficient of dt in expression (346) is equal to zero, and the expression itself reduces to $dp_1, dp_2, \ldots dq_1, dq_2, \ldots$. In other words, the points under consideration occupy exactly the same volume at the end of the interval dt as they occupied at the beginning. Thus their density does not change in any small interval dt, and so must remain unaltered through all time.

In brief, there is no tendency for the representative points to crowd into any special region or regions in the phase-space, a result which was first enunciated by Liouville, and is generally known as Liouville's theorem.

207. Imagine now that any dynamical system is found invariably to possess a certain property (e.g. maximum entropy) after being left to itself for a sufficient time. This might a priori be for either of two reasons: either that the points in the phase-space tend to crowd into those regions of the space for which the property is true, or else that the property is true for the whole of the space. Liouville's theorem excludes the first alternative, so that the second must be the true one. We must now examine this in some detail.

Normal Properties and the Normal State

208. Let P denote any property—as for instance maximum entropy—which the system under consideration may or may not possess. Let us, as before, represent all possible states of the system by a collection of points filling the appropriate phase-space. Let the states in which the system possesses the property P occupy a volume W_1 of the space, while those in which the system does not possess the property P occupy the remaining volume $W - W_1$ of the phase-space. If we choose coordinates and momenta for the system at random, we are in effect choosing a point at random in the phase-space to represent the system. The chance that the system shall possess the property P is accordingly W_1/W.

Let us next examine what is the probability that a system initially selected at random shall have the property P after following out its natural motion for a time t. Imagine the phase-space to be filled with a cloud of representative points, so close together that they may be regarded as forming a continuous fluid, and let these points be distributed initially so that the density of this fluid is uniform. Each of these points has an equal chance of representing the system selected initially at random. Let this cloud of points move for the time t, in accordance with the equations of motion of the system. Liouville's theorem shews that at the end of this time, the fluid will still be of uniform density. Thus after any time t the number of points which represent systems possessing the property P will be a fraction W_1/W of the whole, and therefore the same fraction measures the probability that the system shall possess the property P after time t.

It follows that if the system is found always to possess the property P after a sufficient time has elapsed, this can only be because the ratio $W_1/(W - W_1)$ is infinite.

A property P which is such that the ratio $W_1/(W - W_1)$ is infinite will be called a "normal property" of the system, while a system which possesses all the normal properties of which it is capable will be said to be in the "normal state".

The Normal State

209. So long as a system has only a finite number of coordinates, the ratio $W_1/(W - W_1)$ corresponding to any property P will necessarily have a finite value. But in a gas, the number of degrees of freedom is so large that it may be treated as infinite, so that it is not surprising that this ratio should become infinite or zero, since W_1 and $W - W_1$ will in general be functions of the number of degrees of freedom.

The various properties of the system will, in general, change with the time, some of them perhaps slowly, some more quickly, some with extreme rapidity. Let us suppose that a property P_1 may in general be expected to change in a time comparable with t_1, a property P_2 in a time comparable with t_2, and so on. After a time t which is very large compared with all of the quantities $t_1, t_2, \ldots$, the system will have had ample time to change all its properties. In a statistical sense, the influence of the initial conditions will have disappeared; if the representative point started in peculiar regions of the generalised space, in which any normal property does not hold, it will have had time to move away from these. The system may therefore be expected to possess all the normal properties of which it is capable, and therefore to be in the normal state.

Complications may arise through the system possessing properties which are not capable of change at all, or for which the time of change is infinite or nearly so.

For instance, a system which is perfectly self-contained and subject to no external influence cannot change its angular momentum, so that the property of having any particular angular momentum is one which cannot be acquired or lost with the lapse of time. In the generalised space, one particular value of the angular momentum, namely zero, is far commoner than any other, so that for the property of having zero angular momentum the value of $W_1/(W - W_1)$ is infinite. Yet the having of zero angular momentum, although a normal property of the system, is not to be regarded as an essential of the normal state; it is not required by the definition of the normal state, since it is not a property which the system is capable of acquiring.

On the other hand, if conditions are such that the system can change its angular momentum, then the property of having the normal value for its angular momentum must be regarded as one of the properties of the normal state. For instance, a gas enclosed in a fixed closed vessel can change its angular momentum, and it is easily seen that the possessing of zero angular momentum is one of the normal properties of such a system. Hence in the final state of the system we must expect the angular momentum of the gas to be zero.

One property which can never be changed in a conservative system is that of having a certain value for the energy. Thus in discussing the normal state, we need only consider systems having a specified amount of energy. In the same way, if the system has other quantities or properties which are invariable, account must be taken of this invariability in specifying the normal state. The various complications which may arise in this way are somewhat difficult to discuss in general terms, but are not difficult to treat in particular cases, as the various examples which occur in the present book will shew.

The Normal Partition of Energy

210. We shall first consider the normal properties associated with the partition of energy.

Let the $2n$ coordinates of position and velocity (344) now be denoted by $\theta_1, \theta_2, \ldots \theta_{2n}$, no distinction in notation being made between the two, and let us suppose that the energy E can be divided into separate and distinct parts $E_1, E_2, \ldots$ such that E_1 depends only on one group of coordinates, say $\theta_1, \theta_2, \ldots \theta_s$; E_2 depends only on another group, distinct from the former, $\theta_{s+1}, \theta_{s+2}, \ldots \theta_{s+t}$, and so on. Also let it be supposed that the number $(s, t, \ldots)$ of coordinates in each of these groups is so great that it may be treated as infinite. Such a case would for instance occur if we had oxygen and nitrogen mixed in a closed vessel. The total energy can be regarded as the sum

energy of oxygen

+ energy of nitrogen

+ energy of containing vessel.

Then

$$E = E_1 + E_2 + \ldots = f_1(\theta_1, \theta_2, \ldots \theta_s) + f_2(\theta_{s+1}, \theta_{s+2}, \ldots \theta_{s+t}) + \ldots$$

Let us define the property P as being possessed by the system when the energy is divided between the various groups of co-ordinates in such a way that

E_1 lies within a small range ϵ extending from E_1 to $E_1 + dE_1$, $\quad\Big\}$
E_2 ,,   ,,   ,,   ,,   ,, $\quad E_2$ to $E_2 + dE_2$, etc.

Then the property P holds throughout that part of the phase-space for which

$$f_1(\theta_1, \theta_2, \ldots \theta_s) \text{ lies between } E_1 \text{ and } E_1 + dE_1, \quad\ldots\ldots(347)$$

$$f_2(\theta_{s+1}, \theta_{s+2}, \ldots \theta_{s+t}) \text{ lies between } E_2 \text{ and } E_2 + dE_2, \quad\ldots\ldots(348)$$

and so on.

The volume W_1, throughout which the property holds, may be written in the form

$$W_1 = \iiint \ldots \int d\theta_1 d\theta_2 \ldots d\theta_{2n},$$

where the integration is taken throughout the region defined by the conditions (347), (348), etc. This again may be written in the form of the product

$$W_1 = \left(\iiint \ldots d\theta_1 d\theta_2 \ldots d\theta_s \right) \left(\iiint \ldots d\theta_{s+1} d\theta_{s+2} \ldots d\theta_{s+t} \right) \ldots,$$

$$\ldots\ldots(349)$$

where the first integral is taken within the limits specified by (347), the second integral within the limits (348), and so on. The first integral can depend only on E_1 and dE_1, and so must be of the form $F_1(E_1)\,dE_1$. Similarly the second must be of the form $F_2(E_2)\,dE_2$, and so on. Thus W_1 must be of the form

$$W_1 = F_1(E_1)\,F_2(E_2)\ldots dE_1\,dE_2\ldots = F_1(E_1)\,F_2(E_2)\ldots \epsilon^s.$$

$$\ldots\ldots(350)$$

We shall not attempt to evaluate W_1 directly, but shall attack the simpler problem of finding for what particular values of $E_1, E_2, \ldots$ W_1 has its maximum value.

If we slightly change the partition of energy, by altering the values of $E_1, E_2, \ldots$ to $E_1+\delta E_1, E_2+\delta E_2, \ldots$, but keeping the ranges $dE_1, dE_2, \ldots$ unaltered and each equal to ϵ, the change in $\log W_1$ is given by

$$\delta \log W_1 = \delta \log F_1(E_1) + \delta \log F_2(E_2) + \ldots$$

$$= \frac{d \log F_1(E_1)}{dE_1}\,\delta E_1 + \frac{d \log F_2(E_2)}{dE_2}\,\delta E_2 + \ldots. \quad \ldots\ldots(351)$$

These changes $\delta E_1, \delta E_2, \ldots$ cannot be anything we please; their sum must be zero, because the total energy $E_1 + E_2 + \ldots$ must not change its value. Thus we must have

$$\delta E_1 = -\delta E_2 - \delta E_3 - \delta E_4 - \ldots,$$

and equation (351) becomes

$$\delta \log W_1 = \left\{ \frac{d \log F_2(E_2)}{dE_2} - \frac{d \log F_1(E_1)}{dE_1} \right\} \delta E_2$$

$$+ \left\{ \frac{d \log F_3(E_3)}{dE_3} - \frac{d \log F_1(E_1)}{dE_1} \right\} \delta E_3 + \ldots.$$

If W_1 is to have its maximum value, this must vanish for all values of $\delta E_2, \delta E_3, \ldots$, so that we must have

$$\frac{d \log F_1(E_1)}{dE_1} = \frac{d \log F_2(E_2)}{dE_2} = \frac{d \log F_3(E_3)}{dE_3}. \quad \ldots\ldots(352)$$

The solution of these equations, together with

$$E = E_1 + E_2 + E_3 + \ldots,$$

will give the most probable partition of energy for a system of assigned total energy E.

Equipartition of Energy

211. Next suppose that part at least of the system we have had under discussion consists of the molecules of a gas, these being all similar and n in number. Part of the energy of this system will be the energy of the translatory motion of these n molecules; let us identify this with E_1, so that

$$E_1 = \tfrac{1}{2}m(u_1^2 + v_1^2 + w_1^2 + u_2^2 + v_2^2 + w_2^2 + \ldots),$$

where m is the mass of each molecule, and $u_1, v_1, w_1; u_2, v_2, w_2; \ldots$ are the velocities of individual molecules. We can identify $mu_1, mv_1, mw_1, mu_2, mv_2, mw_2$, etc. with the $\theta_1, \theta_2, \ldots \theta_s$ of equation (349), so that

$$F(E_1)\,dE_1 = \iiint \ldots d\theta_1 d\theta_2 \ldots d\theta_s, \qquad \ldots\ldots(353)$$

where the integral is taken over all values of $\theta_1, \theta_2, \ldots \theta_s$ for which E_1 lies within the specified range of extent dE_1.

If the range for E_1 were from 0 to some assigned value E_1', the integral could easily be evaluated, for the integration would then extend throughout the interior of the boundary

$$u_1^2 + v_1^2 + \ldots = \frac{2E_1'}{m},$$

or
$$\theta_1^2 + \theta_2^2 + \ldots + \theta_s^2 = 2mE_1',$$

which is a sphere in s-dimensional space. The integration merely evaluates the volume of this sphere, and since the radius of this sphere is $(2mE')^{\frac{1}{2}}$, its volume must be $C(2mE')^{\frac{1}{2}s}$, where C is a constant of which the value need not concern us.*

On differentiating this in respect to E_1', we find that the value of the integral taken over all values which make E_1 lie within a range dE_1 is

$$Csm(2mE_1)^{\frac{1}{2}s-1}dE_1.$$

This is precisely the quantity we have called $F(E_1)\,dE_1$, so that, on carrying out the differentiation,

$$\frac{d\log F(E_1)}{dE_1} = \frac{\tfrac{1}{2}s - 1}{E_1}.$$

---

* Actually the value of C is $\dfrac{\pi^{\frac{1}{2}s}}{\Gamma(\tfrac{1}{2}s + 1)}$.

We have already identified s with the number of degrees of freedom of the n molecules to move in space, so that $s = 3n$. If n is a very large number, the difference between $\frac{1}{2}s$ and $\frac{1}{2}s - 1$ may be neglected, and we may write

$$\frac{d \log F(E_1)}{dE_1} = \frac{\frac{1}{2}s}{E_1} = \frac{\frac{3}{2}n}{E_1} = \frac{3}{2\bar{e}_1}, \qquad \ldots\ldots(354)$$

where $\bar{e}_1$ denotes the mean energy of motion $\frac{1}{2}m(u^2+v^2+w^2)$ of a single molecule, of which the value is known to be $\frac{3}{2}RT$. Thus equation (352) assumes the form

$$\frac{d \log F(E_1)}{dE_1} = \frac{d \log F(E_2)}{dE_2} = \ldots = \frac{1}{RT}. \qquad \ldots\ldots(355)$$

Actual molecules may possess energy other than that of this motion in space—as for instance energy of rotation or of vibrations of their internal structure. Often the energies of such motions may be expressed as a sum of squares of the coordinates.

In general, if different parts $E_1, E_2, \ldots$ of the whole energy consist of $s, t, \ldots$ squared terms, we may evaluate the most probable values of E_1, E_2 in precisely the same way as above, and find that, whatever the physical interpretation of the squared terms may be, the most probable partition of energy is given, as regards those parts of the energy for which the energy function is quadratic, by the equations

$$E_1 = \frac{1}{2}sRT, \quad E_2 = \frac{1}{2}tRT, \text{ etc.,} \qquad \ldots\ldots(356)$$

where $s, t, \ldots$ are the number of coordinates concerned in the quadratic functions $E_1, E_2, \ldots$.

It is proved in Appendix III that, as far at least as these parts of the energy are concerned, the partition of energy expressed by equations (356) is not only the most likely partition, but also expresses a "normal" partition in the sense of §208; that is to say, this partition is infinitely more probable than any other.

We have accordingly shewn that those parts of a system, say $E_1, E_2, \ldots$, in which the energy is of quadratic type, will necessarily tend to the partition of energy specified by equations (356). These equations express the Theorem of Equipartition of Energy in its most general form:

The energy to be expected for any part of the total energy which can be expressed as a sum of squares is of amount $\frac{1}{2}RT$ for every squared term in this part of the energy.

The Normal Distribution of Coordinates

212. Not only will there be a normal partition of energy, but there will also be a normal way for the separate coordinates to be arranged so as to give this particular energy. For the simple case of a gas composed of molecules which behave like hard elastic spheres, this has been shewn to be Maxwell's law of distribution which we obtained in the form

$$f(u,\ v,\ w)\,du\,dv\,dw = Ce^{-hm(u^2+v^2+w^2)+\alpha_2 mu+\alpha_3 mv+\alpha_4 mw+\alpha_5}.$$
$$\dots\dots(357)$$

In the present case let us suppose that in addition to its coordinates and velocities in space x, y, z and u, v, w, the molecule contains internal motions expressed by additional coordinates $\phi_1, \phi_2, \phi_3, \dots$. Let us suppose that the energy of a molecule is ϵ_1, this of course being conserved at collisions, and that in addition to this there are other quantities $\epsilon_2, \epsilon_3, \epsilon_4, \dots$, which are conserved at collisions. Then it is shewn in Appendix IV that the normal law of distribution of coordinates is

$$Ce^{-\alpha_1\epsilon_1-\alpha_2\epsilon_2-\alpha_3\epsilon_3-\alpha_4\epsilon_4-\cdots}dx\,dy\,dz\,du\,dv\,dw\,d\phi_1\,d\phi_2\dots \quad \dots\dots(358)$$

If the only quantity which remains constant is the energy, this reduces to

$$Ce^{-2h\epsilon_1}dx\,dy\,dz\,du\,dv\,dw\,d\phi_1\,d\phi_2\dots, \quad \dots\dots(359)$$

where $2h$ replaces α_1 so as to agree with the notation already used, $2h$ being equal to $1/RT$. If certain of the coordinates, say $\phi_1, \phi_2, \dots \phi_s$, enter into the energy ϵ only through their squares, so that the value of ϵ is of the form

$$\epsilon = \tfrac{1}{2}\beta_1\phi_1^2 + \tfrac{1}{2}\beta_2\phi_2^2 + \dots + \tfrac{1}{2}\beta_s\phi_s^2 + \Phi,$$

where Φ does not involve $\phi_1, \phi_2, \dots \phi_s$, but only $\phi_{s+1}, \phi_{s+2}, \dots \phi_{2p}$, then the law of distribution may be written in the form

$$C(e^{-h\beta_1\phi_1^2}d\phi_1)(e^{-h\beta_2\phi_2^2}d\phi_2)\dots(e^{-h\beta_s\phi_s^2}d\phi_s)e^{-2h\Phi}d\phi_{s+1}\dots d\phi_{2p}.$$

This shews that there is no correlation between the distributions of $\phi_1, \phi_2, \dots \phi_s$. The law of distribution of any single coordinate, say ϕ_1, is of the form

$$\sqrt{\frac{h\beta_1}{\pi}}\, e^{-h\beta_1\phi_1^2} d\phi_1, \qquad \dots\dots(360)$$

the constant being determined from the condition that the integral, taken from $\phi = -\infty$ to $\phi = +\infty$, shall be equal to unity.

The mean value of the contribution from ϕ_1 to the energy is of course $\dfrac{1}{4h}$ or $\tfrac{1}{2}RT$, as before, and we have

$$\overline{\tfrac{1}{2}mu^2} = \overline{\tfrac{1}{2}mv^2} = \overline{\tfrac{1}{2}mw^2} = \overline{\tfrac{1}{2}\beta_1\phi_1^2} = \overline{\tfrac{1}{2}\beta_2\phi_2^2} = \dots = \frac{1}{4h} = \tfrac{1}{2}RT,$$

expressing the theorem of equipartition of energy in a new form.

THERMODYNAMICS

213. Let us suppose that our gas, containing vessel, etc. are in their most probable state, specified by the equations

$$\frac{d\log F_1(E_1)}{dE_1} = \frac{d\log F_2(E_2)}{dE_2} = \dots = \frac{1}{RT}. \qquad \dots\dots(361)$$

Let the gas now be heated from outside, and suppose that, after the gas has again attained its most probable state, we find that $E_1, E_2, \dots$ have been increased to $E_1 + dE_1,\ E_2 + dE_2, \dots$.

With W_1 defined by equation (350), let us put

$$P = \log W_1 = \log\left[F_1(E_1)\,F_2(E_2)\dots\right] + \text{a constant.} \qquad \dots\dots(362)$$

Then the heating of the gas changes P by an amount

$$dP = \frac{d\log F_1(E_1)}{dE_1}dE_1 + \frac{d\log F_2(E_2)}{dE_2}dE_2 + \dots.$$

Using relation (361) this becomes

$$dP = (dE_1 + dE_2 + \dots)\frac{1}{RT}$$

$$= \frac{dQ}{RT},$$

where dQ is the total amount of energy that has been added to the gas. Thus

$$\frac{dQ}{T} = R\, dP.$$

The second law of thermodynamics tells us (§ 12) that dQ/T must be a perfect differential; we now see that it is the differential of RP where P is defined by equation (362), or again of $R \log W_1$. According to thermodynamical theory dQ/T is equal to dS, where S is the entropy. We accordingly see that

$$S = RP + \text{a constant} = R \log W_1 + \text{a constant}.$$

We now see that the entropy S is the same thing as P, except for a multiplying constant R, and an additive constant which is of no consequence because we are only interested in changes in S. It is also the same thing as $\log W_1$, the logarithm of the probability that the energy of the system under discussion shall be divided according to the partition

$$E_1, E_2, \ldots.$$

We have already seen that the most probable state is one in which W_1 is a maximum; we may now put this in the form that it is one in which the entropy is a maximum.

To get a clearer idea of the physical meaning of entropy, let us suppose that a system consists of only two parts, of energies E_1, E_2. Let the total energy $E = E_1 + E_2$ remain unaltered, but let a quantity of energy dQ pass from the first part E_1 to the second part E_2, so that the partition of energy is changed from E_1, E_2 to

$$E_1 - dQ, \quad E_2 + dQ.$$

Let the two parts of the system be at different temperatures T_1, T_2. Then the subtraction of energy dQ from the first part of the system reduces the total entropy by dQ/T_1, while the addition of energy dQ to the second part of the system increases the total entropy by dQ/T_2. The net gain to the entropy is accordingly given by

$$dS = dQ\left(\frac{1}{T_2} - \frac{1}{T_1}\right).$$

If the process is one which occurs in the ordinary course of nature, in which heat-energy can flow only from a hot body to a cooler one, T_1 is necessarily greater than T_2. Hence in a natural change dS is positive—the entropy necessarily increases, and a final steady state is only reached when the entropy has attained the maximum value which is possible for it.

Since S is equal to $R \log W_1$, where W_1 measures the probability of a state, this is equivalent to saying that in a natural motion of a system the probability W_1 continually increases, and the system only reaches a final steady state when the probability W_1 has reached its maximum value.

Entropy and Probability

214. So far we have defined and discussed entropy as something associated with a temperature T, but temperature has no meaning except for a system, or part of a system, which is in a steady state. Entropy has in fact only been defined for a system in a steady state.

The results just obtained make it possible to remove this restriction. When parts of the system have specific temperatures attached to them, we have seen that

$$S = R \log W_1 + \text{a constant.} \qquad \ldots\ldots(363)$$

But W_1 has a definite meaning with reference to any partition of energy in the system, so that we may define a more general entropy S by equation (363).

Here W_1 is the volume of that part of the generalised space which represents systems having the partition in question. This volume measures, in a sense, the probability that this special partition shall prevail, so that the entropy is proportional to the logarithm of the probability.

An increase of entropy is thus associated with an increase of probability. The law that the entropy always increases means that a system passes always from a less likely to a more likely state. The law that the entropy is a maximum in the steady state means simply that the steady state is the state of maximum probability.

We have so far thought of the steady state and the final state as the same thing, but we now see that strictly speaking the two are not identical. The value of W, the volume or probability, which corresponds to the steady state is far larger than that which corresponds to any other state. If the number of molecules in the gas were infinite, it would be infinitely larger. With the number of molecules in the gas enormously large, it is enormously larger, but not infinitely so. If W_1 is the probability of the state of maximum entropy, and W_2 that of any other specified state, W_1/W_2 will be very large but not absolutely infinite.

Thus if a gas starts in the second state there is a very large probability that, after a sufficient time, it will be found in the steady state. There is also a finite, although very small, probability that if it starts in the steady state it will be found after a time t in the second state. Thus the increase of entropy is only a probability, not a certainty, although the more we increase the number of molecules in the gas, the nearer this probability comes to unity.

If the system is allowed to undergo its natural changes for an infinite time, it is almost a certainty that at some time it will pass through the second state. This must be so unless the trajectories in the generalised space all avoid the region of volume W_2, and any such avoidance would be contrary to Liouville's theorem.

We may, then, say that an increase of entropy is not a matter of certainty, but only of very high probability. And if the system continues in existence for long enough, it is certain that at some time a decrease of entropy will occur.

215. When applied to concrete instances, these results seem at first sight somewhat startling. To borrow an illustration from Lord Kelvin, if we have a bar of iron initially at uniform temperature, and subject neither to external disturbance nor to loss of energy, it is infinitely probable that, given sufficient time, the temperature of one half will at some time differ by a finite amount from that of the other half. Or again, if we place a vessel full of water over a fire, it is only probable, and not certain, that the water will boil instead of freeze. And moreover, if we attempt to boil the water a sufficient number of times, it is infinitely probable that the water will, on some occasions, freeze instead of boil. The freezing of the water, in this case, does not in any way imply a contravention of the laws of nature: the occurrence is merely what is commonly described as a "coincidence", exactly similar in kind to that which has taken place when the dealer in a game of bridge finds that he has all the cards of one suit in his own hand.

The analogy of the distribution of a pack of cards will help us to see further into the problem presented by the entropy of a gas. In dealing cards, it is just as likely that the dealer will have the thirteen spades as that he will have any other thirteen cards that we like to specify. The occurrence of a hand composed of thirteen spades might, however, be justly regarded as a "coincidence", whereas the occurrence of any specified hand in which the cards were more thoroughly mixed, could not reasonably be so regarded. The explanation is that there are comparatively few ways in which a hand which is all spades can be dealt, but a great number in which a mixed hand can be dealt.

A similar remark applies to the result of putting cold water over a hot fire. There are comparatively few ways in which the fire can get hotter, and the water colder, but a great many ways in which the fire can impart heat to the water—a proposition which becomes obvious on looking at it in terms of the phase-space. Speaking loosely, it is just as likely that the water will freeze as that it will boil in any specified way. There are, however, so many ways in which the water can boil, all these ways being indistinguishable to us, that we can say that it is practically certain that the water will boil.

The increase of entropy, then, simply means the passage from a more easily distinguishable state to a less easily distinguishable state, or, in terms of the generalised space, from a less probable to a more probable configuration.

216. In general, the energy of a dynamical system is the sum of two parts—kinetic and potential. We see from equations (362) and (363) that the entropy consists of two parts, the former depending on the energy of the molecules of the gas, and the latter on their positions. So far we have considered only variations in the first term, resulting from inequalities in the temperature of the gas. Similar remarks could, however, be made about the variations of the second term, these denoting inequalities in the density of the gas. A single illustration, suggested by Willard Gibbs,[*] will, perhaps, make clear what is meant.

If we put red and blue ink together in a vessel, and stir them up, common experience tells us that, if the inks differ initially in nothing more than colour, the result of stirring is a uniform violet

* *Elementary Principles of Statistical Mechanics*, p. 144.

ink. Here we have the passage from a more easily distinguishable to a less easily distinguishable arrangement of coloured inks. If, however, we start by stirring a uniform violet ink composed of a mixture of red and blue inks, then it is possible, although not probable, that the effect of the stirring will be to separate the inks of different colour, so that one half of the vessel is occupied solely by red, and the other solely by blue ink. And from the dynamical standpoint it is no less probable that this should occur, than that we should be able to start stirring inks which were separated initially as regards colour.

217. With reference to this subject, some well-known remarks of Maxwell* are of interest. He says: "One of the best established facts in thermodynamics is that it is impossible in a system en-closed in an envelope which permits neither change of volume nor passage of heat, and in which both the temperature and the pressure are everywhere the same, to produce any inequality of temperature or of pressure without the expenditure of work. This is the second law of thermodynamics, and it is undoubtedly true so long as we can deal with bodies only in mass and have no power of perceiving or handling the separate molecules of which they are made up. But if we conceive a being whose faculties are so sharpened that he can follow every molecule in its course, such a being, whose attributes are still as essentially finite as our own, would be able to do what is at present impossible to us. For we have seen that the molecules in a vessel full of air at uniform temperature are moving with velocities by no means uniform though the mean velocity of any great number of them, arbitrarily selected, is almost exactly uniform. Now let us suppose that such a vessel is divided into two portions A and B, by a division in which there is a small hole, and that a being, who can see the individual molecules, opens and closes this hole, so as to allow only the swifter molecules to pass from A to B, and only the slower ones to pass from B to A. He will thus, without expendi-ture of work, raise the temperature of B and lower that of A, in contradiction to the second law of thermodynamics."

Thus Maxwell's sorting demon could effect in a very short time what would probably take a very long time to come about if left

* *Theory of Heat*, p. 328.

to the play of chance. There would, however, be nothing contrary to natural laws in the one case any more than in the other.

FLUCTUATIONS

218. Formula (358) or (359) shews that when no external forces act on a gas, the normal state is one in which the density is uniform throughout. In this state, and no other, the entropy is a maximum.

Nevertheless, we saw in § 80 how the number of molecules in any small volume of the gas must continually fluctuate as molecules enter or leave the volume. The entropy, then, can never be permanently equal to its maximum value; there must be continual fluctuation below and up to this value, even when a gas is in a completely steady state. The phenomena we mentioned in § 215, as for instance the freezing of a kettle of water put over a fire, result from fluctuations of this kind.

To study these fluctuations in detail, let us fix our attention on a small element of volume Ω in a gas of N molecules, and let Ω be one sth part of the whole volume of the gas. In the state of maximum entropy, the volume Ω will of course contain one sth part of all the molecules, namely, N/s, which we shall denote by n. We proceed to examine how the actual number will fluctuate about its mean value n.

Since all positions in the gas are equally likely for each molecule (§ 82), the chance that any particular molecule A shall be inside Ω at any specified instant is $\dfrac{1}{s}$; the chance of its being outside is $1 - \dfrac{1}{s}$. Thus the chance that a group of p selected molecules $A, B, C, \ldots$ shall all be inside, while the remaining $N - p$ molecules are all outside, is

$$\left(\frac{1}{s}\right)^p \left(1 - \frac{1}{s}\right)^{N-p} .$$

Such a group of p molecules can be selected out of the N molecules of the gas in

$$\frac{N!}{p!\,(N-p)!} = \frac{N(N-1)(N-2)\ldots(N-p+1)}{p!}$$

ways, so that the whole chance Q that there shall be p molecules inside Ω, and the remaining $(N-p)$ outside, is

$$Q = \frac{N(N-1)(N-2)\ldots(N-p+1)}{p!}\left(\frac{1}{s}\right)^p\left(1-\frac{1}{s}\right)^{N-p}.$$

If Ω is only a small fraction of the whole volume of the gas, s will be large, and p small compared with N. In these circumstances, we may replace $N(N-1)\ldots(N-p+1)$ by N^p, and the value of Q becomes

$$Q = \frac{1}{p!}\left(\frac{N}{s}\right)^p e^{\frac{N-p}{s}}$$

$$= \frac{n^p}{p!}\,e^{-n}. \qquad\qquad \ldots\ldots(364)$$

This is the probability that the volume Ω shall contain p molecules in place of its average number n. If p is itself a large number, although still small compared with N, we can express this probability in a simpler form by using Stirling's approximation for $p!$, namely

$$p! = \frac{1}{\sqrt{2\pi p}}\left(\frac{p}{e}\right)^p.$$

Equation (364) now becomes

$$Q = \sqrt{2\pi p}\left(\frac{n}{p}\right)^p e^{p-n}. \qquad\qquad \ldots\ldots(365)$$

Let us now put $p = n + \delta$, so that δ is the excess of molecules in Ω caused by fluctuations about the mean. Taking logarithms of both sides, equation (365) becomes

$$\log Q = \log\sqrt{2\pi p} - (n+\delta)\log\left(1+\frac{\delta}{n}\right) + \delta$$

$$= \log\sqrt{2\pi p} - (n+\delta)\left(\frac{\delta}{n} - \frac{\delta^2}{2n^2} + \ldots\right) + \delta$$

$$= \log\sqrt{2\pi n} - \frac{\delta^2}{2n} + \text{terms of higher order in } \delta,$$

so that

$$Q = \sqrt{2\pi n}\,e^{-\frac{\delta^2}{2n}}. \qquad\qquad \ldots\ldots(366)$$

This gives the chance that the number of molecules in the volume Ω shall differ by δ from the average value n. Or, if we divide up the whole gas into "cells" of volume Ω, it gives the proportion of the whole number in which there are $n + \delta$ molecules.

From the integrals given in Appendix VI (p. 306), we readily find that the mean numerical value of δ is $0.798\sqrt{n}$, while the mean value of δ^2 is exactly n. Thus as n increases, δ increases as $\sqrt{n}$.

219. An average cubic millimetre of air at ordinary pressure contains 2.7×10^{16} molecules, so that the average fluctuation is about 1.3×10^8 molecules—hundreds of millions of molecules in excess or defect which, however, form only 0.0000005 per cent of the whole. At a cathode-ray tube pressure of 0.001 mm., a cubic millimetre contains only about 4×10^{10} molecules, so that the average fluctuation is about 200,000 molecules and there may be appreciable variations of density over lengths comparable with a millimetre. A gas which is near to its critical point is abnormally sensitive to such density variations, which now shew themselves optically, giving rise to a characteristic opalescence. From measurements on this phenomenon also, estimates of Loschmidt's number have been made, ranging from 6×10^{23} to 7.7×10^{23}.[*] While these cannot claim a high degree of accuracy, they suffice to shew that the explanation in terms of fluctuations is correct.

[*] S. E. Virgo, *Science Progress*, **108** (1933), p. 634.

Chapter XI

CALORIMETRY AND MOLECULAR STRUCTURE

Specific Heats

220. A brief account of the problem of calorimetry has already been given in §§ 20, 21 (p. 33). We proceed now to study the problem in more detail.

Calorimetry

Let us suppose that the pressure of a gas is known in terms of its volume and temperature, so that

$$p = f(v, T). \qquad \qquad \ldots\ldots(367)$$

As in § 12, the equation of energy is

$$dQ = Nd\bar{E} + p\,dv. \qquad \qquad \ldots\ldots(368)$$

Specific heat at constant volume. If the volume v is kept constant while the heat dQ is absorbed, we may put $dv = 0$, and the equation of energy becomes

$$dQ = Nd\bar{E}.$$

If C_v is the specific heat at constant volume—i.e. the heat required to raise the temperature of the gas by one degree—then the amount of heat required to raise the mass Nm through a temperature-difference dT will be $C_v Nm\,dT$. Changing to units of energy, the number of ergs needed to raise the temperature by dT will be

$$J C_v Nm\,dT,$$

where J is the mechanical equivalent of heat. This must then be the value of dQ, or of $Nd\bar{E}$. Hence we have, as the value of C_v,

$$C_v = \frac{1}{Jm}\frac{d\bar{E}}{dT}.$$

Specific heat at constant pressure. If the pressure is kept constant while the heat dQ is absorbed, both v and T change, so that we must not put $dv = 0$ in equation (368). There is a change of volume dv which is related to the change of temperature dT

through the restriction that the value of p, as given by equation (367), shall remain unaltered. In fact we have the general formula

$$dp = \left(\frac{\partial p}{\partial v}\right) dv + \left(\frac{\partial p}{\partial T}\right) dT,$$

so that when $dp = 0$,

$$dv = -\frac{\partial p/\partial T}{\partial p/\partial v} dT.$$

Thus when the heat dQ is absorbed at constant pressure ($dp = 0$), the equation of energy becomes

$$dQ = Nd\bar{E} - p\left[\frac{\partial p/\partial T}{\partial p/\partial v}\right] dT.$$

If C_p is the specific heat at constant pressure, this is equal to $JC_p Nm\, dT$, so that

$$C_p = C_v - \frac{p}{JNm}\left[\frac{\partial p/\partial T}{\partial p/\partial v}\right].$$

Specific Heats of a Perfect Gas

221. If the gas is perfect, $p = NRT/v$, whence we readily find that

$$\frac{\partial p/\partial T}{\partial p/\partial v} = -\frac{v}{T} = \frac{NR}{p},$$

and the formulae for the specific heats become

$$C_v = \frac{1}{Jm}\frac{d\bar{E}}{dT}, \qquad \ldots\ldots(369)$$

$$C_p = C_v + \frac{R}{Jm}, \qquad \ldots\ldots(370)$$

the formulae already obtained in § 20.

Molecular Structure

222. Experiment shews that for many gases C_p and C_v are approximately independent of the temperature through a considerable range of temperatures and pressures in which the gas behaves approximately as a perfect gas. This, as is shewn by a reference to equation (369), must mean that $d\bar{E}/dT$ is a constant, and therefore that the mean energy of a molecule of the gas stands in a constant ratio to the translational energy. As in § 21, let us denote this ratio by $(1+\beta)$, so that β is the ratio of internal to translational energy. Then

$$\bar{E} = (1+\beta)\tfrac{3}{2}RT, \qquad \ldots\ldots(371)$$

so that

$$\frac{d\bar{E}}{dT} = \tfrac{3}{2}R(1+\beta). \qquad \ldots\ldots(372)$$

Equations (369) and (370) now become

$$C_v = \tfrac{3}{2}(1+\beta)\frac{R}{Jm}, \qquad \ldots\ldots(373)$$

$$C_p = [1 + \tfrac{3}{2}(1+\beta)]\frac{R}{Jm}, \qquad \ldots\ldots(374)$$

whence, on division,

$$\frac{C_p}{C_v} = 1 + \frac{2}{3(1+\beta)}. \qquad \ldots\ldots(375)$$

The quantities $d\bar{E}/dT$ and β (the two being connected by relation (372)) can only be evaluated when the internal structure of the molecule is known. We have not sufficient knowledge of this internal structure to evaluate these quantities directly, but their values can to some extent be determined from a comparison of the specific heat formulae and the experimentally determined values of the specific heats, and the values obtained in this way provide a basis for the discussion of the structure of molecules.

As an example of this procedure, we may examine the case of air which for the moment, as frequently in the kinetic theory, may be thought of as consisting of similar molecules.

For γ, the ratio of the specific heats, under a pressure of 1 atmosphere, the following values have been obtained by Koch and others:*

Values of γ for air at 760 mm. pressure:

$$\theta = -79\cdot3^\circ \text{ C.,} \qquad \gamma = 1\cdot405,$$
$$\theta = \quad\ 0^\circ \text{ C.,} \qquad \gamma = 1\cdot404,$$
$$\theta = \ 100^\circ \text{ C.,} \qquad \gamma = 1\cdot403,$$
$$\theta = \ 500^\circ \text{ C.,} \qquad \gamma = 1\cdot399,$$
$$\theta = \ 900^\circ \text{ C.,} \qquad \gamma = 1\cdot39.\dagger$$

* Kaye and Laby's *Physical Constants* (1936).
† Value given by Kalähne.

These numbers shew that at this pressure γ is almost independent of the temperature, and approximately equal to $1\frac{2}{5}$. Equation (375), namely

$$\gamma = 1 + \frac{2}{3(1+\beta)}, \qquad \qquad \ldots\ldots(376)$$

now shews that $\beta = \frac{2}{3}$.

At higher pressures the value of γ is by no means constant, as is shewn by the following observations by Koch.*

Values of γ for Air at High Pressures

	$\theta = -79\cdot3°$ C.	$\theta = 0°$ C.
$p =$ 1 atmos.	$\gamma = 1\cdot405$	$\gamma = 1\cdot404$
25 „	1·57	1·47
100 „	2·21	1·66
200 „	2·33	1·85

223. Generally speaking, it is found that for monatomic gases, and for the more permanent diatomic gases, there exists a range of the kind we have found for air, within which the specific heats remain approximately constant. But in the case of more complex gases, it frequently happens that no such range appears to exist.

The table below gives observed values of γ for a number of the more common gases of both types. The third column contains values of β calculated from formula (376), but it will be remembered that these values have no obvious physical meaning except within the range in which the specific heats remain approximately steady.

The figures in this table, in conjunction with a large mass of other experimental evidence, shew that the value of β is approximately equal to zero ($\gamma = 1\frac{2}{3}$) for the monatomic gases—mercury, krypton, argon and helium. It is nearly equal to $\frac{2}{3}$ ($\gamma = 1\frac{2}{5}$) throughout the steady range for a number of diatomic gases—hydrogen, oxygen, carbon-monoxide and others. When we pass to temperatures below the steady range, β is found to decrease with great rapidity.

* *Ann. d. Physik*, **26** (1908), p. 551.

Gas	γ	β	3β	Observer
Monatomic gases				
Helium at 18° C.	1·63	0·06	0·19	Behn and Geiger
Helium at −180°	1·667	0·00	0·00	Scheel and Heuse
Argon at 0°	1·667	0·00	0·00	Niemeyer
Neon at 19°	1·642	0·04	0·13	Ramsay
Krypton at 19°	1·689	−0·04	−0·13	,,
Xenon at 19°	1·666	0·00	0·00	,,
Mercury vapour at 310°	1·666	0·00	0·00	Kundt and Warburg
Diatomic gases				
Hydrogen at 16°	1·407	0·64	1·90	Scheel and Heuse
Hydrogen at −76°	1·453	0·47	1·41	,,
Hydrogen at −181°	1·597	0·12	0·37	,,
Nitrogen at −20°	1·400	0·67	2·00	,,
Nitrogen at −161°	1·468	0·42	1·27	,,
Oxygen at 20°	1·399	0·67	2·01	,,
Oxygen at −76°	1·416	0·60	1·81	,,
Oxygen at −181°	1·447	0·49	1·48	,,
Nitric oxide	1·394	0·69	2·08	Masson
Chlorine	1·333	1·00	3·0	Strecker
Bromine	1·293	1·27	3·8	,,
Iodine	1·293	1·27	3·8	,,
Hydrochloric acid (HCl)	1·40	0·67	2·0	,,
Bromine iodide (Br I)	1·33	1·00	3·0	,,
Chlorine iodide (Cl I)	1·317	1·10	3·3	,,
Polyatomic gases				
Water vapour	1·305	1·19	3·56	Makower
Carbon-dioxide	1·300	1·22	3·67	Lummer and Pringsheim
Nitrogen peroxide (NO₂ at 150°)	1·31	1·15	3·45	Natanson
Nitrous oxide (N₂O)	1·324	1·06	3·18	Leduc

The fourth column gives the value of 3β, and we notice a tendency for these values to cluster round integral values, except at low temperatures. These values are usually 0 for monatomic gases, 2 and 3 or 4 for diatomic gases; values near to $3\beta = 1$ are conspicuously absent. General dynamical theory gives some indication of what this may mean.

224. The energy of a molecule will consist always of three squared terms representing the kinetic energy of motion, to which may be added any number of other terms representing energy of rotation, of internal vibration, etc.

If these latter terms are n in number, the average value of each is $\frac{1}{2}RT$, which is also the average value of each squared term in the kinetic energy of motion. Thus

$$\beta = \frac{\frac{1}{2}nRT}{\frac{3}{2}RT} = \frac{n}{3},$$

so that $n = 3\beta$. This is the number tabulated in the fourth column on p. 279.

This shews why 3β, which represents the number of degrees of freedom, ought always to be integral. The difficulty now is to understand how it is possible for n to have any other than integral values.

Let us examine different kinds of gas in turn.

Monatomic Gases

225. The six monatomic gases in the table—mercury, xenon, krypton, argon, neon and helium—all have very approximately $\gamma = 1\frac{2}{3}$, $\beta = 0$, and $n = 0$. For these gases, then, there is no appreciable amount of molecular energy except that of translation. This seems to indicate that the molecules of these gases must behave at collision like hard spherical bodies. If they did not do so, an appreciable fraction of the molecular energy of translation would be transformed into some other form of energy at each collision. In these gases the molecule is of course identical with the atom.

Although the atoms of these substances behave like hard spherical bodies at collision, there is abundant evidence that they have a highly complicated internal structure. The helium atom for instance is known to consist of three parts—a positive nucleus, which is identical with the α-particle of radioactivity, and two negative electrons. The helium atom made up in this way must, as a matter of geometry, have six degrees of freedom in addition to its three degrees of freedom of motion in space.

The explanation of why the specific heats of such an atom could be accurately obtained by taking $\beta = 0$ in formulae (373) and (374) presented for many years a problem of the utmost gravity. It is now generally accepted that no satisfactory explanation can be given in terms of the classical system of dynamics. In recent years the new system quantum-dynamics has provided an explanation not only of the behaviour of the monatomic molecule but of many other problems of specific heats in addition. This is, however, outside the range of the present book.

After $n = 0$, the next value theoretically possible would be $n = 1$, giving $\gamma = 1\frac{1}{2}$. No gas is known for which n and γ have these values, even approximately. This provides additional confirmation of the truth of the kinetic theory, since no molecular system could conceivably have its energy expressible as the sum of four squared terms; either a rotation or a vibration would necessarily introduce at least two squared terms into the energy, beyond the three terms representing motion in space. Thus a molecule for which $n = 1$ is an impossibility, and the value $\gamma = 1\frac{1}{2}$ cannot be expected to occur for any type of gas.

Diatomic Gases

226. After $n = 0$, then, the next value which is theoretically possible is $n = 2$, giving $\gamma = 1\frac{2}{5}$. The table shews that n and γ have very approximately these values for hydrogen, nitrogen and oxygen at ordinary temperatures. This indicates two squared terms in the energy which may either represent a vibration of the two atoms relative to one another along the line joining them, or a possibility of a rotation. We must notice that in general the energy of rotation of a rigid body will be represented by three squared terms, but the energy of rotation of a body symmetrical about an axis may be represented by only two, no rotation about the axis of symmetry being set up by collisions.

Whatever is the true physical origin of these two terms, the table on p. 279 shews that their energy falls off rapidly as the temperature falls, particularly in the case of hydrogen. Eucken and others* have found experimentally that as the absolute zero of temperature is approached, the molecules of diatomic gases tend to lose all energy except that of translation, and so behave like the molecules of monatomic gases, with the values $\beta = 0$ and $\gamma = 1\frac{2}{3}$.

* Eucken, *Sitzungsber. Berlin Akad. d. Wissensch.* **6** (1912), p. 141; Scheel and Heuse, *Ann. d. Phys.* **40** (1913), p. 473; Schreiner, *Zeits. f. Phys. Chem.* **112** (1924), p. 1.

More complex Gases

227. It is difficult to discover any law or regularity in the values of n and γ for more complex gases. Various attempts have been made to connect the values of n and γ with the number of atoms in the molecule, but it is now clear that no general law can be expected to relate γ with the number of atoms, independently of the nature of the atoms. For instance, Capstick* found the following values for the methane derivatives:

		γ	$n+3$
Methane	CH_4	1·313	6·4
Methyl chloride	CH_3Cl	1·279	7·2
Methylene chloride	CH_2Cl_2	1·219	9·0
Chloroform	$CHCl_3$	1·154	13·0
Carbon tetrachloride	CCl_4	1·130	15·4

We see that the introduction of the series of chlorine atoms increases n very perceptibly at every step, without increasing the number of atoms in the molecule. Capstick found a similar law for paraffin derivatives; the second chlorine atom introduced into the molecule produced a large change, although the first may or may not have done so.

A similar result was obtained by Strecker,† who found that hydrochloric, hydrobromic, hydriodic acids all have approximately the same values as hydrogen, namely

$$\gamma = 1\cdot4, \qquad n+3 = 5,$$

while for chlorine, bromine and iodine, the values are approximately

Chlorine $\qquad \gamma = 1\cdot333 \quad n+3 = 6,$

Bromine, Iodine $\gamma = 1\cdot293 \quad n+3 = 6\cdot8.$

Similarly for the iodides of bromine and chlorine,

Bromine iodide $\quad \gamma = 1\cdot33 \qquad n+3 = 6,$

Chlorine iodide $\quad \gamma = 1\cdot317 \quad n+3 = 6\cdot3.$

These figures shew that one halogen can be put in the place of hydrogen without producing any difference in the values of γ

---

* *Phil. Trans.* **186** (1895), p. 564; **185** (1894), p. 1.

† Wiedemann's *Annalen*, **13** (1881), p. 20 and **17** (1882), p. 85.

and n, but the substitution of the second halogen atom causes a marked increase in n.

MOLECULAR AGGREGATION

228. The discussion of the physical properties of gases given in this and the preceding chapters has been based upon the supposition that a gas can be regarded as a collection of separate dynamical systems, namely molecules, each of which retains its identity through all time. As a close to this discussion, we may examine what changes are to be expected if the supposition is regarded as an approximation to the truth only, and not as being wholly true. We shall first consider what complications are introduced by the possibilities of molecular aggregation, leaving the discussion of the converse process of dissociation until later.

We again simplify the problem by regarding molecules as point-centres of force, acting on one another with a force depending only on their distance apart. The chance of finding a free molecule of class A inside an element of volume $dx\,dy\,dz$ is now

$$Ae^{-hmc^2}du\,dv\,dw\,dx\,dy\,dz, \qquad \ldots\ldots(377)$$

while the chance of finding two molecules of classes A and B in adjacent elements $dx\,dy\,dz$ and $dx'dy'dz'$ is

$$A^2e^{-hm(c^2+c'^2)-2h\Psi}du\,dv\,dw\,dx\,dy\,dz\,du'\,dv'\,dw'\,dx'\,dy'\,dz',$$

where Ψ is the mutual potential energy of the two molecules.

If we replace the element $dx'\,dy'\,dz'$ by a spherical shell of radii r and $r+dr$ surrounding the centre of the first molecule, this last expression becomes

$$A^2e^{-hm(c^2+c'^2)-2h\Psi}du\,dv\,dw\,du'\,dv'\,dw'\,4\pi r^2\,dr\,dx\,dy\,dz,$$

Ψ being a function of r. If we use the transformations of §109,

$$\mathbf{u} = \tfrac{1}{2}(u+u'), \text{ etc.}, \quad \alpha = u'-u, \text{ etc.},$$

and write

$$\mathbf{u}^2+\mathbf{v}^2+\mathbf{w}^2 = \mathbf{c}^2, \quad \alpha^2+\beta^2+\gamma^2 = V^2,$$

we can transform the foregoing expression into

$$A^2e^{-2hm\mathbf{c}^2}d\mathbf{u}\,d\mathbf{v}\,d\mathbf{w}\,dx\,dy\,dz\,e^{-\frac{1}{2}hmV^2-2h\Psi}d\alpha\,d\beta\,d\gamma\,4\pi r^2\,dr.$$

$$\ldots\ldots(378)$$

The first factor after the A^2 expresses the law of distribution of translational velocities for a double molecule. It is of course the same as if each double molecule were a permanent structure of mass $2m$. The remaining factors express the distribution of those coordinates which may be regarded as internal to the double molecule.

229. So long as the motion of a double molecule is undisturbed by collisions, c^2 remains constant, so that the energy equation shews that $\frac{1}{2}mV^2 + 2\Psi$ remains constant. The orbits which the component molecules can describe about their common centre of gravity fall into two classes, according as they pass to infinity or not. Analytically these two classes are differentiated by the sign of $\frac{1}{2}mV^2 + 2\Psi$. Double molecules for which $\frac{1}{2}mV^2 + 2\Psi$ is positive consist of two molecules which have approached one another from outside each other's sphere of action, and which after passing once within a certain minimum distance of each other, will again recede out of each other's sphere of influence. On the other hand, double molecules for which $\frac{1}{2}mV^2 + 2\Psi$ is negative consist of two molecules describing orbits about one another, these orbits being entirely within the two spheres of action, and this motion continues except in so far as it is interrupted by collisions with other molecules. Clearly double molecules of the first kind are simply pairs of molecules in ordinary collision. In discussing molecular aggregation we must confine our attention to double molecules of the second kind, i.e. those for which $\frac{1}{2}mV^2 + 2\Psi$ is negative. Such double molecules cannot be produced solely by the meeting of two single molecules. It is necessary that while the single molecules are in collision something should happen to change the motion—in fact to change the sign of $\frac{1}{2}mV^2 + 2\Psi$. This might be effected by collision with a third molecule, or possibly if $\frac{1}{2}mV^2 + 2\Psi$ were small at the beginning of an encounter, sufficient energy might be dissipated by radiation for $\frac{1}{2}mV^2 + 2\Psi$ to become negative before the termination of the encounter. We may disregard the consideration of this second possibility for the present, with the remark that if this were the primary cause of aggregation, the equations with which we have been working would not be valid, since they rest upon the assumption of conservation of energy for the molecules alone.

Integrating expression (377) over all values of u, v and w, we find that ν_1, the molecular density of uncombined molecules, is given by

$$\nu_1 = A\left(\frac{\pi}{hm}\right)^{\frac{3}{2}}. \qquad \ldots\ldots(379)$$

Similarly, by integration of expression (378), we find for ν_2 the molecular density of double molecules,

$$\nu_2 = A^2\left(\frac{\pi}{2hm}\right)^{\frac{3}{2}}\iiiint e^{-h[\frac{1}{2}mV^2+2\Psi]}d\alpha\,d\beta\,d\gamma\,4\pi r^2 dr$$

$$= A^2\left(\frac{\pi}{2hm}\right)^{\frac{3}{2}}\iint e^{-h[\frac{1}{2}mV^2+2\Psi]}16\pi^2 V^2 r^2 dV\,dr, \quad \ldots\ldots(380)$$

in which the integration extends over all values of V and r for which $\frac{1}{2}mV^2 + 2\psi$ is negative.

The total number of constituent molecules per unit volume is

$$\nu = \nu_1 + 2\nu_2 + 3\nu_3 + \ldots$$

$$= \nu_1\left(1 + \frac{A}{\sqrt{2}}\iint e^{-h[\frac{1}{2}mV^2+2\Psi]}16\pi^2 V^2 r^2 dV\,dr + A^2(\ldots) + \ldots\right),$$

$$\ldots\ldots(381)$$

so that if q denotes the fraction of the whole mass which is free,

$$q = \frac{\nu_1}{\nu} = \frac{1}{1 + \dfrac{A}{\sqrt{2}}\displaystyle\iint e^{-h[\frac{1}{2}mV^2+2\Psi]}16\pi^2 V^2 r^2 dV dr + \ldots}. \qquad \ldots\ldots(382)$$

Eliminating A from equations (379), (380), etc., we obtain a series of relations of the form

$$\left.\begin{array}{l}\nu_2 = \nu_1^2\phi(T)\\ \nu_3 = \nu_1^3\psi(T),\ \text{etc.}\end{array}\right\}, \qquad \ldots\ldots(383)$$

where ϕ, ψ, $\ldots$ are functions of the temperature only. Equations of this type have formed the basis of practically every theory of aggregation and dissociation.*

---

* Compare, for instance, Boltzmann's Theory, *Wied. Ann.* **32**, p. 39, or *Vorlesungen über Gastheorie*, **2**, § 63; Natanson's Theory, *Wied. Ann.* **38**, p. 288, or Winkelmann's *Handbuch d. Physik*, 3, p. 725, or the theory of J. J. Thomson, *Phil. Mag.* [5] **18** (1884), p. 233. These theories are based on widely different physical assumptions, but all lead to equations of the same general form as (403). The difference of the physical assumptions made shews itself in the different forms for the functions $\phi(T)$, etc.

To study the variation of aggregation with temperature a knowledge of the exact form of the functions $\phi(T)$, $\psi(T)$, etc. is necessary, but it is possible to examine the dependence of aggregation on density without this knowledge.

Dependence of Aggregation on Density

230. For a number of substances, it is probable that no greater degree of aggregation need be considered than that implied in the formation of double molecules. For such substances

$$\nu = \nu_1 + 2\nu_2.$$

Neglecting the Van der Waals corrections, the pressure is given by (equation (56))

$$p = RT(\nu_1 + \nu_2) = \tfrac{1}{2}RT(\nu + \nu_1) = \tfrac{1}{2}R\nu T(1+q), \quad \ldots\ldots(384)$$

where q is introduced from equation (382). Thus it appears that q, the fraction of the whole mass which is free, can be readily obtained from readings of pressure and temperature.

The following table gives the values of $1-q$ calculated in this way from the observations of Natanson* on the density of peroxide of nitrogen:

Aggregation of NO_2

Temp.	Value of $1-q = \dfrac{2\nu_2}{\nu}$			
	$p = 115$ mm.	$p = 250$ mm.	$p = 580$ mm.	$p = 760$ mm.
$\theta = -12\cdot6°$ C.	0·919	—	—	—
$\theta = \quad 0°$	0·837	0·901	—	—
$\theta = \quad 21°$	—	—	0·824	—
$\theta = \quad 49\cdot7°$	0·253	0·370	0·550	—
$\theta = \quad 73\cdot7°$	0·084	0·149	0·263	—
$\theta = \quad 99\cdot8°$	0·031	0·050	0·093	0·117
$\theta = \quad 151\cdot4°$		Inappreciable		

Here the single molecule is NO_2, the double molecule is N_2O_4, and more complex structures are supposed to occur only in negligible amounts. The value of $1-q$ is $2\nu_2/\nu$, and so measures the proportion by mass which occurs in the form N_2O_4.

* Recueil de constantes physiques, p. 168.

Equation (383) predicts that the ratio of ν_2 to ν_1^2 ought to be the same for all readings at the same temperature. We have from this equation

$$\frac{1}{q} = \frac{\nu}{\nu_1} = \frac{\nu_1 + 2\nu_2}{\nu_1} = 1 + 2\nu_1\phi(T),$$

so that

$$\frac{1}{q}\left(\frac{1}{q} - 1\right) = \left(\frac{\nu}{\nu_1}\right)[2\nu_1\phi(T)] = 2\nu\phi(T).$$

Substituting the value of ν given by equation (384), we obtain

$$1 - q^2 = 4pq^2\frac{\phi(T)}{RT},$$

shewing that the ratio $\dfrac{1 - q^2}{pq^2}$ ought to be the same for all readings at the same temperature.

The following table, calculated from the observations in the table of p. 286, will shew to what extent this prediction of theory is borne out by experiment:

Aggregation of NO_2

Temp.	Value of $\dfrac{1-q^2}{pq^2} = \dfrac{4\phi(T)}{RT}$			
	$p = 115$ mm.	$p = 250$ mm.	$p = 580$ mm.	$p = 760$ mm.
$\theta = 49 \cdot 7°$ C.	0·689	0·608	0·680	—
$\theta = 73 \cdot 7°$	0·167	0·152	0·145	—
$\theta = 99 \cdot 8°$	0·056	0·043	0·037	0·037

Dependence of Aggregation on Temperature

231. The degree to which the aggregation varies with temperature depends on the functions $\phi(T)$, $\psi(T)$, etc. introduced in equations (383). From equations (379) and (380), we find

$$\phi(T) = \frac{\nu_2}{\nu_1^2} = \left(\frac{hm}{2\pi}\right)^{\frac{3}{2}}\iint e^{-h[\frac{1}{2}mV^2 + 2\Psi]}16\pi^2V^2r^2\,dV\,dr. \quad \ldots\ldots(385)$$

At the highest temperatures (h very small) the value of $\phi(T)$ will be clearly insignificant, but the presence of the exponential in formula (385) suggests that after $\phi(T)$ has once become appreciable, it must be expected to increase rapidly with falling temperature.

Our knowledge of the structure of matter is not sufficient to enable us to evaluate $\phi(T)$, as given by equation (385), with precision. Progress can only be made by the introduction of simple hypotheses as to the interaction of molecules, which may prove to lead to results near to the truth.

Boltzmann* imagined that potential energy exists between two molecules only when the centre of the second lies within a small and clearly defined region which is of course fixed relative to the first, and that when the second molecule has its centre within this "sensitive region", the potential energy has always the same value Ψ. This region does not necessarily consist of a spherical shell, but if ω denotes its total volume, equation (385) may be written in the form

$$\phi(T) = \left(\frac{hm}{2\pi}\right)^{\frac{3}{2}} \omega e^{-2h\Psi} \int e^{-\frac{1}{2}hmV^2} 4\pi V^2 dV,$$

where ω replaces the integral $4\pi \int r^2 dr$, which has represented the extent of the "sensitive region" in our analysis. If we replace $\frac{1}{2}hmV^2$ by x^2, this equation may be expressed in the form

$$\phi(T) = \frac{4\omega}{\sqrt{\pi}} e^{-2h\Psi} \int_0^{\xi} e^{-x^2} x^2 dx. \qquad \ldots\ldots(386)$$

The upper limit of integration is determined by the condition that $\frac{1}{2}mV^2 + 2\Psi$ shall vanish, and is therefore given by $\xi^2 = -2h\Psi$, the value of Ψ being necessarily negative. If we put $-\Psi = R\beta$, so that β is positive, the value of ξ^2 is $2hR\beta$ or β/T.

For some substances Ψ may be so large that a good approximation can be obtained by taking the integral in equation (386) between the limits $x = 0$ to $x = \infty$. The integration is now readily effected, and we find

$$\phi(T) = \omega e^{-2h\Psi} = \omega e^{-\Psi/RT} = \omega e^{\beta/T},$$

in which Ψ is negative. The degree of dissociation is then given by

$$\frac{1}{q^2} = 1 + \frac{4p\omega}{RT} e^{\beta/T}, \qquad \ldots\ldots(387)$$

* *Vorlesungen über Gastheorie*, **2**, Chap. VI.

which is Boltzmann's formula for molecular aggregation and dissociation.

An almost exactly identical treatment of the problem was given by Willard Gibbs.*

The following table contains the densities of peroxide of nitrogen observed at various temperatures by Deville and Troost,† the pressure being one atmosphere throughout, and also the values calculated from equation (387).

Aggregation of NO_2

Temp. ° C.	Density (observed)	Density (calc.)	Temp. ° C.	Density (observed)	Density (calc.)
183·2	1·57	1·592	80·6	1·80	1·801
154·0	1·58	1·597	70·0	1·92	1·920
135·0	1·60	1·607	60·2	2·08	2·067
121·5	1·62	1·622	49·6	2·27	2·256
111·3	1·65	1·641	39·8	2·46	2·443
100·1	1·68	1·676	35·4	2·53	2·524
90·0	1·72	1·728	26·7	2·65	2·676

Continuity of Liquid and Gaseous States

232. At very high temperatures, the series (381) reduces to its first term, so that $q = 1$, and there are no molecules in permanent combination.

At lower temperatures h is greater, so that not only is A greater, but the exponential $e^{-h[\frac{1}{2}mV^2+2\Psi]}$, in which it will be remembered that the index is always positive, is also greater. The relative importance of the later terms of the series (381) is therefore greater. Finally, we reach values of the temperature for which h has so great a value that the series (381) becomes divergent. According to our analysis the molecules tend to form into clusters at this point, each containing an infinitely great number of molecules, and, ultimately, into one big cluster absorbing all the molecules. By the time this stage is reached the analysis has ceased to apply, as the assumption that the molecular clusters are small, made in § 230, is now invalidated. It is, however, easy to

* *Trans. Connecticut Acad.* **3** (1875), p. 108 and (1877), p. 343; *Silliman Journal*, **18** (1879), p. 277. Also *Coll. Works*, **1**, pp. 55 and 372.

† *Comptes Rendus*, **64** (1867), p. 237.

give a physical interpretation of the point now reached: obviously it is the point at which liquefaction begins, and the collection of molecular clusters is a saturated vapour.

The series (381) may be regarded as a power series in ascending powers of A. Thus for a given value of h, say h_0, there is a single value of A, say A_0, such that the series is convergent for all values of A less than A_0 and is divergent for all values of A greater than A_0. In other words, corresponding to a given temperature, there is a definite density at which the substance liquefies. This of course is the vapour-density corresponding to this temperature. As h increases, A will clearly decrease, and conversely, so that an increase of pressure is accompanied by a rise in the boiling-point of the substance.

Since A depends on ν, the relation between corresponding values h_0, A_0 which has just been obtained may be expressed in the form

$$f(\nu,\ T) = 0, \qquad\qquad \dots\dots(388)$$

expressing the relation between ν and T at the boiling-point of a liquid.

The Critical Point

233. For very small values of h, ν_1 and ν become identical, so that the series (381) cannot become divergent. Thus for very high values of T equation (385) can have no root corresponding to a physically possible state. If T_c is the lowest value of T for which equation (385) has a root, then T_c will be a temperature above which liquefaction cannot possibly set in, no matter how great the density of the gas; in other words, T_c is the critical temperature.

Ordinary algebraic theory tells us that there must be two coincident values of ν given by equation (385) to correspond to the critical temperature T_c, agreeing with what is already known as to the slope of the isothermals at the critical point.

Pressure, Density and Temperature

234. It will now be clear that when a gas or vapour is at a temperature which is only slightly greater than its boiling-point at the pressure in question, it cannot be regarded as consisting of single molecules, but must be supposed to consist partly of single molecules and partly of clusters of two, three or more molecules. If m is the mass of a single molecule, and if $\nu_1, \nu_2, \nu_3, \ldots$ have the same meaning as before, the density is given by

$$\rho = m(\nu_1 + 2\nu_2 + 3\nu_3 + \ldots).$$

In calculating the pressure, each type of cluster must be treated as a separate kind of gas, exerting its own partial pressure. We accordingly obtain for the pressure, as in § 230,

$$p = \frac{1}{2h}(\nu_1 + \nu_2 + \nu_3 + \ldots) = \frac{\rho RT}{m}\left(\frac{\nu_1 + \nu_2 + \nu_3 + \ldots}{\nu_1 + 2\nu_2 + 3\nu_3 + \ldots}\right).$$

From a comparison of this equation with equation (383), remembering that $\nu_1, \nu_2, \ldots$ are functions of T and ρ, it is clear that neither Boyle's law, Charles' law nor Avogadro's law will be satisfied with any accuracy.

235. The observed deviations from the laws obeyed by a perfect gas must of course be attributed partly to aggregation, as has just been explained, and partly to the causes which have already been discussed in Chap. III. The two sets of causes are not, however, altogether independent; so that it is not sufficient to consider the effects separately, and then add. The state of the question is, perhaps, best regarded as follows.

The effect of the forces of cohesion is too complex for an exact mathematical treatment to be possible. We have therefore, in Chap. III and the present chapter, examined their effect with the help of two separate simplifying assumptions. In Chap. III, following Van der Waals, we regarded the gas as a single molecular cluster containing an infinite number of molecules; and in replacing the whole system of the forces of cohesion by a permanent *average* force, we virtually neglected the effect of any formations of small clusters inside the large cluster. In the present chapter, on the other hand, we have been concerned solely with the formation of small clusters, and have disregarded the large cluster altogether. The former treatment, because it failed to take account of the formation of small clusters, led to the erroneous result (equation (71) and § 45) that the internal pressure is proportional to the temperature; the treatment of the present chapter, because it fails to consider the clustering of the gas as a whole, has led to the erroneous conclusion that the internal pressure is identical with the boundary pressure. The situation may then be summed up by saying that the treatment of Chap. III considers only the tendency to *mass-clustering*, while that of the present chapter considers only the tendency to *molecular-clustering*.

So long as the deviations from the behaviour of a perfect gas are small, the two tendencies may be considered separately, and the total deviation regarded as the sum of the two deviations caused by these tendencies separately. On the other hand, as we approach the critical point the phenomena of mass-clustering and molecular-clustering merge into one another, ultimately becoming identical at the critical point. The two effects are no longer additive, for each has become identical with the whole effect.

It must be borne in mind that we have only found an exact mathematical treatment of either effect to be possible by making the assumption that the effect itself is small. In other words, so far as our results apply, the effects are additive. It may be noticed that the deviations from the laws of a perfect gas, which were discussed in Chap. III, fell off proportionally to $1/T$ and $1/T^2$, whereas the deviations discussed in the present chapter vary much more rapidly with the temperature.

Calorimetry

236. The formulae which have been obtained for the specific heats will be affected to a greater or lesser degree by the possibilities of molecular aggregation. For in raising the temperature of the gas work is done not only in increasing the energy of the various molecules, but also in separating a number of molecules from one another's attractions. This latter work will involve an addition to the values of C_p and C_v such as was not contemplated in the earlier analysis of §§ 220–223. We should therefore expect the values of C_p and C_v to be in excess of the values obtained from our earlier formulae, throughout all regions of pressure and temperature in which molecular aggregation can come into play. For instance, the specific heats of nitrogen peroxide have been studied by Berthelot and Ogier,* who give the following values for C_p:

From 27° to   67° C., $C_p = 1.62$,
,,   27° to 100°,   1.46,
,,   27° to 150°,   1.115,
,,   27° to 200°,   0.85,
,,   27° to 300°,   0.64.

The excess in the values of C_p at the low temperatures may be reasonably attributed to the work required to separate molecules of N_2O_4 into pairs of molecules of NO_2.

Steam provides a further illustration of a somewhat different nature. Wet steam is steam in which large molecular clusters occur, dry steam is steam in which the molecules are all separate, and our quantity q measures what engineers speak of as the dryness of wet steam. For the value of γ for wet (saturated) steam, Rankine and Zeuner give respectively the values 1.0625, 1.0646. The value for dry steam is about 1.30. If we used the formula

$$\gamma = 1 + \frac{2}{3(1 + \beta)}$$

for the calculation of n, we should come to the conclusion that $3 + 3\beta$ had the value 32 for wet steam, and 6.6 for dry steam.

The large value of 3β in the former case is fully in keeping with the existence of large clusters of molecules, so large that each has about 32 degrees of freedom.

DISSOCIATION

237. A treatment similar to the foregoing may be applied to the problem of dissociation. The former molecules must be replaced by atoms, and the former clusters of molecules by single molecules.

Let us consider a gas in which the complete molecules are each composed of two atoms, distinguished by the suffixes 1, 2. As in

---

* *Bull. Soc. Chimie* [2], **37** (1882), p. 434; *Comptes Rendus*, **93** (1882), p. 916; *Ann. d. Chim. et de Phys.* [5], **30** (1883), p. 382; *Recueil de constantes physiques*, p. 108.

equations (399) and (400) the laws of distribution of dissociated atoms and complete molecules are

$$\tau_1 = Ae^{-2hE_1},$$
$$\tau_2 = Be^{-2hE_2},$$
$$\tau_{12} = ABe^{-2h(E_1+E_2+\varPsi)},$$

where $\varPsi$ is the potential energy of the two atoms forming the molecule. The analysis will be simplified, and the theory sufficiently illustrated, by regarding the atoms as point centres of force, of masses m_1, m_2 respectively. Thus we obtain as the laws of distribution of velocities for the dissociated atoms

$$\left. \begin{array}{c} Ae^{-hm_1c^2}du\,dv\,dw \\ Be^{-hm_2c^2}du\,dv\,dw \end{array} \right\}, \qquad \ldots\ldots(389)$$

and for the complete molecules

$$ABe^{-h(m_1+m_2)c^2}du\,dv\,dw\,e^{-h\frac{m_1m_2}{m_1+m_2}V^2-2h\varPsi}\,d\alpha\,d\beta\,d\gamma\,4\pi r^2 dr.$$
$$\ldots\ldots(390)$$

This latter law is arrived at in the same way as the law (378), except that the scheme of transformation of velocities must now be taken to be

$$\mathbf{u} = \frac{m_1u+m_2u'}{m_1+m_2}, \quad \alpha = u'-u, \text{ etc.,}$$

this being the transformation already used in § 109.

238. Although the mathematical analysis is similar to that of the aggregation problem, there is an important difference in the physical conditions. The law of distribution (390) is limited to values of the variable such that

$$\frac{m_1m_2}{m_1+m_2}V^2+2\varPsi$$

is negative; as soon as this quantity becomes positive the molecule splits up into its component atoms. Now in the case of molecular aggregation, the attraction between complete molecules is not great, so that $\varPsi$ is a *small* negative quantity, and the range of values for V is correspondingly small. In the case of chemical dissociation $\varPsi$ is a *large* negative quantity, and the range for V is practically unlimited.

An estimate of the value of Ψ can be formed from the amount of heat evolved when chemical combination takes place. For instance, when 2 grammes of hydrogen combine with 16 grammes of oxygen to form 18 grammes of water, the amount of heat developed according to Thomsen's determination is 68,376 units— sufficient to raise the temperature of the whole mass of water by $3600°$ C. The value of V necessary for dissociation to occur is therefore comparable with the mean value of V at $3600°$ C., and these high values of V will be very rare in a gas at ordinary temperatures. The exclusion from the law of distribution (410) of high values of V will therefore have but little effect either on the law of distribution or on the energy represented by the internal degrees of freedom, and we may, without serious error, regard the law of distribution as holding for all values of V.

In such a case, it appears that the molecule may be treated exactly as an ordinary diatomic molecule, supposed incapable of dissociation, but possessing six degrees of freedom, three translational degrees represented by the differentials $du\,dv\,dw$, and three internal degrees represented by the differentials $d\alpha\,d\beta\,d\gamma$.

Since there are six degrees of freedom, the value of γ will be as low as $1\frac{1}{3}$ if potential energy is neglected, and will be even less if potential energy be taken into account. We have, however, seen that for diatomic molecules γ is fairly uniformly equal to $1\frac{2}{5}$, and this shews that the ordinary diatomic molecule must not be treated as consisting of two atoms describing orbits in the way we have imagined.

We are here brought back to the difficulties which have already been encountered in § 225 in connection with the specific heats of gases. The solution of these difficulties is not provided by the old classical dynamics but by the new quantum dynamics. We accordingly leave the question at this stage; it passes out of the scope of the kinetic theory, and so beyond the range of the present book.

Appendix I

MAXWELL'S PROOF OF THE LAW OF DISTRIBUTION OF VELOCITIES

As already mentioned, Maxwell's original proof of the law of distribution of velocities was given in a paper he communicated to the British Association in 1859. Except for a slight change of notation, the form in which it was given is as follows.*

"Let N be the whole number of particles. Let u, v, w be the components of the velocity of each particle in three rectangular directions, and let the number of particles for which u lies between u and $u + du$ be $Nf(u)\,du$, where $f(u)$ is a function of u to be determined.

"The number of particles for which v lies between v and $v + dv$ will be $Nf(v)\,dv$, and the number for which w lies between w and $w + dw$ will be $Nf(w)\,dw$, where f always stands for the same function.

"Now the existence of the velocity u does not in any way affect that of the velocities v or w, since these are all at right angles to each other and independent, so that the number of particles whose velocity lies between u and $u + du$, and also between v and $v + dv$ and also between w and $w + dw$ is

$$Nf(u)\,f(v)\,f(w)\,du\,dv\,dw.$$

If we suppose the N particles to start from the origin at the same instant, then this will be the number in the element of volume $du\,dv\,dw$ after unit of time, and the number referred to unit of volume will be

$$Nf(u)\,f(v)\,f(w).$$

"But the directions of the coordinates are perfectly arbitrary, and therefore this number must depend on the distance from the origin alone, that is

$$f(u)\,f(v)\,f(w) = \phi(u^2 + v^2 + w^2).$$

Solving this functional equation, we find

$$f(u) = Ce^{Au^2}, \quad \phi(u^2 + v^2 + w^2) = C^3 e^{A(u^2 + v^2 + w^2)}.\text{"}$$

This proof is now generally admitted to be unsatisfactory, because it *assumes* the three velocity components to be independent. The velocities do not, however, enter independently into the dynamical equations of collisions between molecules, so that until the contrary has been proved, we should expect to find correlation between these velocities.

* J. C. Maxwell, *Collected Works*, **1**, p. 380.

Appendix II

THE H-THEOREM

In Chap. IV (§ 87) we discovered certain steady states for a gas, but it was not proved that a gas would necessarily attain to one of these steady states after a sufficient time. The following proof of this was originally given by Boltzmann and Lorentz.

With the notation already used in Chap. IV, let H be defined by

$$H = \iiint f \log f \, du \, dv \, dw, \qquad \qquad \text{......}(a)$$

in which the integration extends over all possible values of u, v, w. Thus H is a pure quantity and not a function of u, v, w; it depends solely upon the law of distribution of velocities and therefore remains unchanged so long as this law remains unchanged. Hence a necessary condition for a steady state is given by $dH/dt = 0$. We proceed to evaluate dH/dt in the general case.

After an interval dt, the value of $f \log f$ corresponding to any specified values of u, v, w will of course have changed into

$$f \log f + \frac{\partial}{\partial t} (f \log f) \, dt,$$

or, what is the same thing, into

$$f \log f + (1 + \log f) \frac{\partial f}{\partial t} \, dt.$$

Hence the increase in H, which may be written $\frac{dH}{dt} \, dt$, will be given by

$$\frac{dH}{dt} \, dt = \left\{ \iiint (1 + \log f) \frac{\partial f}{\partial t} \, du \, dv \, dw \right\} dt, \qquad \text{......}(b)$$

or, substituting the value of $\partial f / \partial t$ from equation (306),

$$\frac{dH}{dt} = \nu \iiint\!\!\iiint\!\!\int (1 + \log f) \, (\bar{f}\bar{f}' - ff') V \sigma^2 \cos \theta \, du \, dv \, dw \, du' \, dv' \, dw' \, d\omega. \qquad \text{......}(c)$$

Since H depends on molecules of all classes $(A, B, C, ...)$ equally, formula (c) necessarily remains true if we interchange the rôles of molecules of classes A and B. We then have the same equation, except that, as the first factor inside the integral, $1 + \log f'$ replaces $1 + \log f$. Adding together these two values for dH/dt, we obtain

$$2 \frac{dH}{dt} = \nu \iiint\!\!\iiint\!\!\int (2 + \log ff') \, (\bar{f}\bar{f}' - ff') V \sigma^2 \cos \theta \, du \, dv \, dw \, du' \, dv' \, dw' \, d\omega. \qquad \text{......}(d)$$

This equation expresses dH/dt as the sum of a number of contributions, one from every possible class of collision. The typical class of collision is taken to be class α, in which

$$u,\ v,\ w,\ u',\ v',\ w'$$

become changed into

$$\bar{u},\ \bar{v},\ \bar{w},\ \bar{u}',\ \bar{v}',\ \bar{w}'.$$

If we use the same equation, but take as the typical collision one of class β, in which

$$\bar{u},\ \bar{v},\ \bar{w},\ \bar{u}',\ \bar{v}',\ \bar{w}'$$

become changed into

$$u,\ v,\ w,\ u',\ v',\ w',$$

we obtain, as a still different form for dH/dt,

$$2\frac{dH}{dt} = \nu \iiint\!\!\iiint\!\!\iiint (2+\log \bar{f}\bar{f}')\,(ff'-\bar{f}\bar{f}')V\sigma^2 \cos\theta\ d\bar{u}\,d\bar{v}\,d\bar{w}\,d\bar{u}'\,d\bar{v}'\,d\bar{w}'\,d\omega.$$
$$\dots\dots(e)$$

As in § 86, we may replace the product of the first six differentials on the right hand of this equation by $du\,dv\,dw\,du'\,dv'\,dw'$, and if we add this modified value of dH/dt to that given by equation (d), we obtain

$$4\frac{dH}{dt} = \nu \iiint\!\!\iiint\!\!\int\!\!\int (\log ff' - \log \bar{f}\bar{f}')(\bar{f}\bar{f}'-ff')V\sigma^2 \cos\theta\ du\,dv\,dw\,du'\,dv'\,dw'\,d\omega.$$
$$\dots\dots(f)$$

Now $(\log ff' - \log \bar{f}\bar{f}')$ is positive or negative according as ff' is greater or is less than $\bar{f}\bar{f}'$ and is therefore always of the sign opposite to that of $\bar{f}\bar{f}'-ff$. Hence the product

$$(\log ff' - \log \bar{f}\bar{f}')\,(\bar{f}\bar{f}'-ff'),$$

if not zero, is necessarily negative. Since $V\cos\theta$, the relative velocity along the line of centres before impact, is necessarily positive for every type of collision, it follows that the integrand of equation (f) is always either negative or zero. Hence equation (f) shews that dH/dt is either negative or zero.

Hence as the motion of the gas progresses, H continually decreases until it reaches a final state in which $dH/dt = 0$. We have seen that every contribution to dH/dt on the right of equation (f) is negative or zero. Thus in the final steady state, every contribution is zero. In other words, we must have

$$ff' = \bar{f}\bar{f}'$$

for every type of collision, bringing us back exactly to the solutions we discussed in § 87.

Appendix III

THE NORMAL PARTITION OF ENERGY

In §§ 210, 211 we found that the most probable partition of energy $(E_1, E_2, E_3, ...)$ was given by the equations (355)

$$\frac{d \log F_1(E_1)}{dE_1} = \frac{d \log F_2(E_2)}{dE_2} = \frac{d \log F_3(E_3)}{dE_3} = \frac{1}{RT}, \qquad ......(a)$$

where $E_1 + E_2 + E_3 + ... = E$, the total energy of the system.

In the most general partition of energy, let us write, as in § 211,

$$P = \log F_1(E_1) + \log F_2(E_2) + \log F_3(E_3) + ...$$

For any partition of energy $E_1 + \epsilon_1, E_2 + \epsilon_2, E_3 + \epsilon_3, ...,$ adjacent to the most probable, the value of P is

$$P = \Sigma \log F_1(E_1 + \epsilon_1)$$
$$= \Sigma \log F_1(E_1) + \Sigma \epsilon_1 \frac{d \log F_1(E_1)}{dE_1} + \tfrac{1}{2} \Sigma \epsilon_1^2 \frac{d^2 \log F_1(E_1)}{dE_1^2} + ... \qquad ......(b)$$

If the total energy remains the same, we have $\Sigma \epsilon_1 = 0$, so that the second term on the right vanishes, and the equation reduces to

$$P = P_0 + \tfrac{1}{2} \Sigma \epsilon_1^2 \frac{d^2 \log F_1(E_1)}{dE_1^2} + ..., \qquad ......(c)$$

where P_0 is the value of P for the most probable partition of energy.

In the special case in which $E_1, E_2, ...$ consist of $s, t, ...$ squared terms, this assumes the form

$$P - P_0 = - \Sigma \frac{\epsilon_1^2}{s(RT)^2} + \tfrac{4}{3} \Sigma \frac{\epsilon_1^3}{s^2(RT)^3} - .... \qquad ......(d)$$

It has already been seen that the only stationary value of P is given by $\epsilon_1 = \epsilon_2 = ... = 0$. This makes $P = P_0$, and an inspection of the right hand of equation (d) shews that this value is a true maximum.

As we recede from the value $\epsilon_1 = \epsilon_2 = ... = 0$, it is clear that $P - P_0$ becomes finite as soon as ϵ_1 becomes comparable with $\sqrt{s}.RT$, ϵ_2 with $\sqrt{t}.RT$, and so on. For such values of $\epsilon_1, \epsilon_2, ...$ the first term on the right of equation (d) is infinitely greater than any of the succeeding terms, and the value of $P - P_0$ reduces to

$$P - P_0 = - \Sigma \frac{\epsilon_1^2}{s(RT)^2}. \qquad ......(e)$$

For values of $\epsilon_1, \epsilon_2, ...$ greater than these $P - P_0$ becomes equal to $-\infty$

From equations (350) and (362), the general value of W_1 is

$$W_1 = e^P dE_1 dE_2 ..., \qquad ......(f)$$

while the whole value of $W_1 + W_2$ will be

$$W_1 + W_2 = \iint ... e^P d\epsilon_1 d\epsilon_2 .... \qquad ......(g)$$

Thus the whole value of the integral (g) comes from a small range of values surrounding the values $\epsilon_1 = \epsilon_2 = \ldots = 0$; i.e. the values of $E_1, E_2, \ldots$ given by equations (a). Thus the integral (g) reduces to the right-hand member of equation (f), the small range dE_1 being comparable with $\sqrt{s}.RT$, the small range dE_2 being comparable with $\sqrt{t}.RT$, and so on. These small ranges are of course small in comparison with the whole values of $E_1, E_2, \ldots$; thus dE_1 is comparable with $E_1/\sqrt{s}$, dE_2 with $E_2/\sqrt{t}$, and so on.

With such values for the small ranges $dE_1, dE_2, \ldots$, the value of $W_1 + W_2$ given by equation (g) becomes identical with the value of W_1 given by equation (f). Thus we have W_1/W_2 infinite, shewing that the partition of energy now under consideration is a normal property of the system.

The proof that the remaining parts of the system, if any, in which the energy is not of this type, will necessarily tend to the partition of energy given by equations (e) is more difficult, since the sign of the terms on the right of equation (c) must necessarily be a matter of uncertainty so long as the form of the energy-function remains unspecified. It is, however, clear that the arrangement of the loci $E_1 = \text{cons.}$, $E_2 = \text{cons.}$, etc., in the phase-space must, in every case, be of the same general type as that in the simple case just considered, from which we may infer that $P - P_0$ must, in the more general case also, be of negative sign. It again follows that W_1/W_2 must be infinite, so that the most probable partition of energy as expressed by equations (a) is now seen to be a normal property of the system.

We accordingly see that every system must pass to a final state in which W_1, and therefore also the entropy, is a maximum. In this way we obtain an analytical proof of the second law of thermodynamics, which may now be regarded as being on a mathematical, instead of on a purely empirical basis.

Appendix IV

THE LAW OF DISTRIBUTION OF COORDINATES

We fix our attention on a part of a dynamical structure, this part consisting of N similar units, which we may think of as molecules for definiteness, each unit possessing p degrees of freedom, and therefore having its state specified by $2p$ quantities $\phi_1, \phi_2, \ldots \phi_{2p}$, these being co-ordinates of position and their corresponding momenta, as in § 205.

Imagine a generalised space of $2p$ dimensions constructed, having

$$\phi_1, \phi_2, \ldots \phi_{2p}$$

as orthogonal coordinates. Then the state of any molecule of the system can be represented by a single point in this space, namely the point whose coordinates are equal to the coordinates $\phi_1, \phi_2, \ldots \phi_{2p}$ specifying the state (i.e. velocity and positional coordinates) of the molecule. The states of all the molecules can be represented by a collection of points in this space, one point for each molecule. We want to find the law according to which these points are distributed in the space.

Let τ denote the density of points in this space—the quantity which we are trying to find—so that

$$\tau\, d\phi_1 d\phi_2 \ldots d\phi_{2p} \qquad \qquad \ldots\ldots(a)$$

will be the number of points (or molecules) such that ϕ_1 lies between ϕ_1 and $\phi_1 + d\phi_1$, ϕ_2 between ϕ_2 and $\phi_2 + d\phi_2$, and so on.

Let us now suppose the whole of this space divided up into n small rectangular elements of volume, each of equal size ω, and let these be identified by numbers 1, 2, 3, ... Let us fix our attention on a special distribution of points, which is such that the number of points in elements 1, 2, 3, ... are respectively $a_1, a_2, a_3, \ldots$. Let any distribution of points giving these particular numbers $a_1, a_2, a_3, \ldots$ be spoken of as a distribution of class A. Similarly any distribution of points giving another set of numbers $b_1, b_2, b_3, \ldots$ may be spoken of as a distribution of class B, and so on.

Each point in the original generalised space will correspond, as in § 206, to a complete distribution of points in the space now under consideration. The distribution corresponding to some of these original points will be a distribution of points of class A, corresponding to others it will be a distribution of class B, and so on. We proceed to evaluate the volume, say W_A, of the original generalised space which is such that the points in it represent systems for which the distribution of coordinates is of class A.

This volume is readily seen to be given by

$$W_A = \frac{N!}{a_1! a_2! a_3! \ldots}\, \omega^N \iiint \ldots d\chi_1 d\chi_2 d\chi_3 \ldots \qquad \ldots\ldots(b)$$

In this expression the first factor on the right hand represents the number of ways in which it is possible to distribute the N points representing the N different molecules, between the n different elements, subject only to the condition of the final arrangement being of type A. The remaining factor, say λ, given by

$$\lambda = \omega^N \int\int\int \ldots d\chi_1 d\chi_2 d\chi_3 \ldots \qquad \ldots\ldots(c)$$

represents the volume of the generalised space which corresponds to each one of these arrangements, $\chi_1, \chi_2, \chi_3, \ldots$ being coordinates of parts of the system other than the N molecules under consideration. If we write

$$\theta_A = \frac{N!}{a_1! a_2! a_3! \ldots}, \qquad \ldots\ldots(d)$$

and use a similar notation for a system of class B, etc., then the volumes W_A, W_B, ... are given by

$$W_A = \lambda \theta_A, \quad W_B = \lambda \theta_B, \text{ etc.} \qquad \ldots\ldots(e)$$

According to the well-known theorem of Stirling, the value of $p!$ when p is very large approximates to the value

$$\underset{p=\infty}{\text{Lt}}\ p! = \sqrt{2p\pi} \left(\frac{p}{e}\right)^p.$$

On taking logarithms of both sides, this becomes

$$\underset{p=\infty}{\text{Lt}}\ \log p! = \tfrac{1}{2} \log 2\pi + (p + \tfrac{1}{2}) \log p - p.$$

Taking logarithms of both sides of equation (d), and using this value for $\log p!$,

$$\log \theta_A = \log N! - \overset{s=n}{\underset{s=1}{\Sigma}} \log a_s!$$

$$= (N + \tfrac{1}{2}) \log N - \tfrac{1}{2}(n-1) \log 2\pi - \overset{s=n}{\underset{s=1}{\Sigma}} (a_s + \tfrac{1}{2}) \log a_s.$$

If we put

$$K_a = \frac{1}{N} \overset{s=n}{\underset{s=1}{\Sigma}} (a_s + \tfrac{1}{2}) \log \frac{na_s}{N}, \qquad \ldots\ldots(f)$$

this gives as the value of θ_A

$$\theta_A = n^{N + \frac{1}{2}n} (2\pi N)^{-\frac{1}{2}(n-1)} e^{-NK_a}. \qquad \ldots\ldots(g)$$

Since $W_A = \lambda \theta_A$, etc., it is clear that W_A is proportional to e^{-NK_a}.

The most probable partition of energy is obviously obtained by making W_A a maximum, and therefore K a minimum, for different values of $a_1, a_2, \ldots$. For the variation of K_a, as given by equation (f), we find

$$\delta K = \frac{1}{N} \overset{s=n}{\underset{s=1}{\Sigma}} \left\{ \log \frac{na_s}{N} + 1 + \frac{1}{2a_s} \right\} \delta a_s. \qquad \ldots\ldots(h)$$

The variations $\delta a_1, \delta a_2, \ldots$ are not independent. They are necessarily connected by two relations, and in some cases by more. Of the two relations which are certain, the first expresses that the total number of molecules remains equal to the prescribed number N, and is therefore expressed by the equation

$$\sum_{s=1}^{s=n} \delta a_s = 0, \qquad \qquad \ldots\ldots(i)$$

while the second relation expresses that the total energy of the N molecules is equal to the allotted amount E_1. Let ϵ_1 denote the energy associated with a molecule represented by a point in cell 1, so that ϵ_1 is a function of $\phi_1, \phi_2, \ldots \phi_{2p}$, the coordinates of the first cell. Let ϵ_2 be the energy associated with a molecule represented by a point in cell 2, and so on. Then the total energy of the N molecules, when the distribution of coordinates is of class A, will clearly be $a_1\epsilon_1 + a_2\epsilon_2 + \ldots$, so that we must have

$$\sum_{s=1}^{s=n} \epsilon_s a_s = E_1,$$

and on variation of this we obtain the relation

$$\sum_{s=1}^{s=n} \epsilon_s \delta a_s = 0. \qquad \qquad \ldots\ldots(j)$$

Any other relations there may be will be derived from equations of similar type, expressing that the total of some quantity μ summed over all the molecules will have an assigned value. The integral equation will be of the form

$$\sum_{s=1}^{s=n} \mu_s a_s = M,$$

and the corresponding relation between the quantities $\delta a_1, \delta a_2, \ldots$ will be

$$\sum_{s=1}^{s=n} \mu_s \delta a_s = 0. \qquad \qquad \ldots\ldots(k)$$

In many problems there will be six equations of this type, the different μ's representing three components of linear momentum, and three components of angular momentum. We may, however, be content to take one relation as typical of all, and shall suppose it to be given by the equations just written down.

Following a well-known procedure, we now multiply equations (i), (j), (k) by undetermined multipliers p, q, r, and add corresponding members of these equations and equation (h). We obtain

$$\delta K = \sum_{s=1}^{s=n} \left\{ 1 + \frac{1}{2a_s} + \log \frac{na_s}{N} + p + q\epsilon_s + r\mu_s \right\} \delta a_s,$$

and the maximum value of K is now given by the equations

$$1 + \frac{1}{2a_s} + \log \frac{na_s}{N} + p + q\epsilon_s + r\mu_s = 0, \quad (s = 1, 2, \ldots n).$$

Since $a_1, a_2, \ldots$ are all supposed to be large quantities, the term $\dfrac{1}{2a_s}$ may be neglected, and we obtain

$$\frac{na_s}{N} = e^{-(1+p)} e^{-(q\epsilon_s + r\mu_s)}. \qquad \qquad \ldots\ldots(l)$$

If τ is the quantity defined in expression (a), the value of a_s will be $\tau\omega$, where τ refers to the sth cell. Equation (l) becomes

$$\tau = \frac{N}{n\omega} e^{-(1+p)} e^{-(q\epsilon + r\mu)},$$

which is true for all values of $\phi_1', \phi_2, \ldots \phi_{2p}$, since equation (l) was true for every cell. Changing the constants, this equation may be rewritten

$$\tau = C e^{-2h\epsilon} e^{-(r_1\mu_1 + r_2\mu_2 + \cdots)},$$

in which the one typical quantity μ is now replaced by the actual series of quantities $\mu_1, \mu_2, \ldots$. Using this value for τ, the law of distribution of coordinates (cf. expression (a)) is seen to be given by

$$C e^{-2h\epsilon} e^{-(r_1\mu_1 + r_2\mu_2 + \cdots)} d\phi_1 d\phi_2 \ldots d\phi_{2p}.$$

From the result obtained in Appendix III, it follows that this distribution of co-ordinates is infinitely more probable than any other, and so expresses a normal state of the system.

If there is no potential energy, there will be three co-ordinates x, y, z fixing the position of the molecule in space, and these will not occur except through the differential $dx\,dy\,dz$. Thus the chance of finding a molecule in a small element of volume $dx\,dy\,dz$, however chosen, will be proportional simply to $dx\,dy\,dz$. This provides the justification for the assumption of molecular chaos introduced in § 82.

Appendix V

TABLES FOR NUMERICAL CALCULATIONS

The following tables will be found of use for the various numerical calculations which are likely to be needed in connection with kinetic theory problems. The values of $\psi(x)$ are from a table by Tait in the paper already referred to (p. 145).

x	x^2	e^{-x^2}	$\dfrac{2}{\sqrt{\pi}}\displaystyle\int_0^x e^{-x^2}\,dx$	$\psi(x)$ Defined by equation (159), p. 140
0·1	0·01	0·99005	0·11246	0·20066
0·2	0·04	0·96080	0·22270	0·40531
0·3	0·09	0·91393	0·32863	0·61784
0·4	0·16	0·85214	0·42839	0·84200
0·5	0·25	0·77880	0·52050	1·08132
0·6	0·36	0·69768	0·60386	1·33907
0·7	0·49	0·61263	0·67780	1·61819
0·8	0·64	0·52729	0·74210	1·92132
0·9	0·81	0·44486	0·79691	2·25072
1·0	1·00	0·36788	0·84270	2·60835
1·1	1·21	0·29820	0·88021	2·99582
1·2	1·44	0·23693	0·91031	3·41448
1·3	1·69	0·18452	0·93401	3·86538
1·4	1·96	0·14086	0·95229	4·34939
1·5	2·25	0·10540	0·96611	4·86713
1·6	2·56	0·07730	0·97635	5·41911
1·7	2·89	0·05558	0·98379	6·00570
1·8	3·24	0·03916	0·98909	6·62715
1·9	3·61	0·02705	0·99279	7·28366
2·0	4·00	0·01832	0·99532	7·97536
2·1	4·41	0·01215	0·99702	8·70234
2·2	4·84	0·00791	0·99814	9·46467
2·3	5·29	0·00504	0·99886	10·26236
2·4	5·76	0·00315	0·99931	11·09547
2·5	6·25	0·00193	0·99959	11·96402
2·6	6·76	0·00116	0·99976	12·86798
2·7	7·29	0·00068	0·99987	13·80734
2·8	7·84	0·00039	0·99992	14·78225
2·9	8·41	0·00022	0·99996	15·79255
3·0	9·00	0·00012	0·99998	16·83830

Appendix VI

INTEGRALS INVOLVING EXPONENTIALS

In making Kinetic Theory calculations, we are frequently confronted with integrals of the type

$$\int u^n e^{-\lambda u^2} du, \qquad \qquad \ldots\ldots\text{(i)}$$

where n is a whole member. Such an integral can be evaluated in finite terms when n is odd, and can be made to depend on the integral

$$\int_0^u e^{-\lambda u^2} du, \qquad \qquad \ldots\ldots\text{(ii)}$$

when n is even. In each case the reduction is most quickly performed by successive integrations by parts with respect to u^2. Tables for the evaluation of the integral (ii) have already been given in Appendix v.

When the limits of integration are from $u = 0$ to $u = \infty$, the results of integration are expressed by the formulae

$$\int_0^\infty u^{2\kappa} e^{-\lambda u^2} du = \frac{1\,.\,3\ldots(2\kappa-1)}{2^{\kappa+1}} \sqrt{\frac{\pi}{\lambda^{2\kappa+1}}},$$

$$\int_0^\infty u^{2\kappa+1} e^{-\lambda u^2} du = \frac{\kappa!}{2\lambda^{\kappa+1}}.$$

The following cases of the general formulae are of such frequent occurrence that it may be useful to give the results separately:

$$\int_0^\infty e^{-hmu^2} du = \frac{1}{2}\sqrt{\frac{\pi}{hm}}, \qquad \int_0^\infty e^{-hmu^2} u\, du = \frac{1}{2hm},$$

$$\int_0^\infty e^{-hmu^2} u^2 du = \frac{1}{4}\sqrt{\frac{\pi}{h^3 m^3}}, \qquad \int_0^\infty e^{-hmu^2} u^3 du = \frac{1}{2h^2 m^2},$$

$$\int_0^\infty e^{-hmu^2} u^4 du = \frac{3}{8}\sqrt{\frac{\pi}{h^5 m^5}}, \qquad \int_0^\infty e^{-hmu^2} u^5 du = \frac{1}{h^3 m^3},$$

$$\int_0^\infty e^{-hmu^2} u^6 du = \frac{15}{16}\sqrt{\frac{\pi}{h^7 m^7}}, \qquad \int_0^\infty e^{-hmu^2} u^7 du = \frac{3}{h^4 m^4}.$$

Each integral can be obtained by differentiating the one immediately above it with respect to hm. In this way the system can be extended indefinitely.

INDEX OF SUBJECTS

INDEX OF NAMES